BARRY COMMONER'S CONTRIBUTION TO THE ENVIRONMENTAL MOVEMENT: SCIENCE AND SOCIAL ACTION

Edited by
Dr. David Kriebel
with the Assistance of Mary Lee Dunn

Work, Health and Environment Series
Series Editors: Charles Levenstein and John Wooding

LONDON AND NEW YORK

First published 2002 by Baywood Publishing Company, Inc.

2 Park Square, Milton Park, Abingdon, Oxfordshire OX14 4RN
52 Vanderbilt Avenue, New York, NY 10017

Routledge is an imprint of the Taylor & Francis Group, an informa business

First issued in hardback 2019

Copyright © 2002 Taylor & Francis

All rights reserved. No part of this book may be reprinted or reproduced or utilised in any form or by any electronic, mechanical, or other means, now known or hereafter invented, including photocopying and recording, or in any information storage or retrieval system, without permission in writing from the publishers.

Notice:
Product or corporate names may be trademarks or registered trademarks, and are used only for identification and explanation without intent to infringe.

Library of Congress Cataloging-in-Publication Data

Barry Commoner's contribution to the environmental movement / edited by David Kriebel.
    p. cm. - - (Work, health and environment series)
  Includes bibliographical references and index.
  ISBN 0-89503-219-8 (paper)
  1. Commoner, Barry, 1917- -Congresses. 2. Environmentalism- -United
States- -Congresses. I. Kriebel, David L. II. Series.

GE197 .B37 2000
363.7'0525'0973- -dc21   00-041425

Library of Congress Catalog Number: 00-041425

ISBN 13: 978-0-89503-219-5 (pbk)
ISBN 13: 978-0-415-78565-5 (hbk)

 Printed in the United Kingdom
by Henry Ling Limited

Contents

Introduction 1

CHAPTER 1 5
Barry Commoner: The Father of Grass-Roots Environmentalism
Peter Montague

CHAPTER 2 15
The Day Before Yesterday: The Committees for Nuclear and
Environmental Information
Virginia Warner Brodine

CHAPTER 3 25
Crossing Paths: Science and the Working Class
Tony Mazzocchi

CHAPTER 4 31
Real Junk Science: The Corruption of Science by Corporate Money
Ralph Nader

CHAPTER 5 45
Barry Commoner's Science: An Anecdotal Overview
Danny Kohl

CHAPTER 6 57
Barry Commoner and the Hamburger Story: Can Ideology Prevail
Over Science?
Piero Dolara

CHAPTER 7 63
The Contribution of Barry Commoner to the Renewal of the Italian Left
Giovanni Berlinguer

CHAPTER 8 67
Barry Commoner's Day
Chicco Testa

CHAPTER 9 73
What is Yet to Be Done
Barry Commoner

Contributors 87

Index 89

Introduction

Few people have made greater contributions to protecting and improving the environment than the scientist, teacher, author, activist Dr. Barry Commoner. For half a century, Dr. Commoner has been an international leader in the environmental movement. On the occasion of his eightieth birthday, in May 1997, a symposium was held at Cooper Union in New York City at which a series of invited speakers discussed Dr. Commoner's contributions to a wide range of environmental issues and campaigns. In the pages that follow, a selection of those invited talks are published.

Dr. Commoner was born in 1917 in Brooklyn, the son of Russian immigrants. After attending Columbia University, he obtained masters and doctoral degrees in botany from Harvard. He served in the U.S. Navy from 1942 to 1946, and in 1947 began his academic career as an associate professor of plant physiology at Washington University in St. Louis, where he remained for more than thirty years. In 1966, he founded the Center for the Biology of Natural Systems (CBNS) at Washington University, one of the first scientific research centers founded explicitly to study environmental problems. The Center was from the start multi-disciplinary, or as Dr. Commoner insists, *a-disciplinary,* because of the now well-accepted idea that environmental problems do not respect traditional academic boundaries.

Barry Commoner's writing and speaking have had tremendous influence. His 1972 book *The Closing Circle* (Knopf) won for him the Phi Beta Kappa award, and the International Prize for Safeguarding the Environment from the city of Cervia, Italy. This book remains essential reading, laying out the central tenets of a holistic view of the impacts of humans and technology on the environment. Two subsequent books, *The Poverty of Power* (Knopf, 1976) and *The Politics of Energy* (Knopf, 1979), have also won literary awards.

Since the 1950s, Barry Commoner has played a pivotal role in nearly every important phase of the environmental movement, including opposition to nuclear weapons testing in the 1950s, the science information movement of the 1960s, the energy debates of the 1970s, and on through pioneering research on organic farming, recycling, and toxic chemical substitution in recent years. Dr. Commoner saw early on that arguments for dangerous technologies (nuclear power, synthetic fertilizers, pesticides, incinerators) were being wrapped in scientific jargon to deflect public scrutiny. He championed the idea of empowering citizens by providing them with accurate and useful information. He also realized that environmental pollution is created by production, and that eliminating pollution requires fundamental changes in the ways things are made.

Several themes emerge from the contributions which follow, themes which are highly relevant and even critical for the future successes of the environmental movement. These themes include:

- Environmental pollution is created by production, and eliminating pollution requires fundamental changes in systems of production;
- Scientists have a social responsibility to make their work relevant and understandable to the public;
- Sound political strategy in the environmental field must be based on sound science; and
- Sound science is *a-disciplinary* science—traditional academic disciplines prevent the kind of "systems thinking" that leads to truly fundamental solutions to social problems.

We hope this volume will stimulate thought and discussion on the lessons of the recent past for the future directions of the environmental movement. In the final chapter, Dr. Commoner argues forcefully that there is much that remains to be done.

The organizing committee for the symposium is grateful for the generous support of the following organizations and individuals:

The Jennifer Altman Foundation
The Ben & Jerry's Foundation
Blue Mountain Center
The Nathan Cummings Foundation
The Joyce Foundation
The Jessie Smith Noyes Foundation
Stan and Iris Ovshinsky
Queens College, City University of New York
The Turner Foundation

The Organizing Committee:

Sharon Peyser, CBNS
Mark Cohen, CBNS
David Kriebel, Department of Work Environment,
    University of Massachusetts Lowell
Ken Geiser, Department of Work Environment,
    University of Massachusetts Lowell

Margaret M. Quinn, Department of Work Environment,
   University of Massachusetts Lowell
Tom Webster, Department of Environmental Health,
   Boston University School of Public Health
Piero Dolara, Department of Pharmacology,
   Universita' degli Studi, Florence, Italy

CHAPTER 1
Barry Commoner: The Father of Grass-Roots Environmentalism

Peter Montague

In the late 1950s and early 1960s, Barry Commoner and his colleagues in the Committee for Nuclear Information and at Washington University in St. Louis developed many of the fundamental arguments and ideas that, today, underpin and propel the grassroots movement for environmental justice, such as: moral wisdom resides in the citizenry; scientists have no special wisdom in moral matters; scientists must make alliances with citizens; pollution must be prevented because it cannot be managed; the burden of proof rests on the polluter; citizens have a right to know; the principle of precautionary action should guide our decisions; environmental impact assessments are essential; and risk assessment is political.

Where did the grass-roots environmental movement come from? Most people would say it began with action by citizens in Warren County, North Carolina, or in Niagara Falls, New York in the late 1970s and early 1980s—and of course they would be partly right. But the world was prepared for grass-roots environmentalism because of certain ideas, and to a surprising degree those ideas originated with one person—a scientist working in St. Louis, Missouri, named Barry Commoner.

In the late 1950s and early 1960s, Commoner developed many of the fundamental ideas that today propel the burgeoning movement of grass-roots environmental activism. Specifically, in Commoner's early writings, I find the following ideas that, today, seem entirely contemporary and widely accepted:

- Moral wisdom resides in the citizenry
- Scientists have no special competence in moral matters
- Scientists have an obligation to make alliances with citizens

- Pollution must be prevented; it cannot be successfully managed
- The burden of proof rests properly on the polluter
- Citizens have a right to know
- The precautionary principle should guide our decisions
- Environmental impact assessment is a necessary tool
- Risk assessment is political, not scientific

Commoner was born May 28, 1917, in Brooklyn, New York, the son of a Russian immigrant tailor. In 1933 he entered Columbia University, then Harvard, earning a Ph.D. in cellular biology in 1941. During World War II, he served in the Naval Air Force. In 1947 he took a faculty position with Washington University in St. Louis where he soon distinguished himself as an exceptionally creative and insightful researcher, studying viruses and elusive "free radicals" in living tissues.

Commoner continued to publish work on proteins and free radicals for twenty years, but in the early 1950s something happened that caught his attention and turned his interest to larger questions. On the morning of April 25, 1953, a nuclear bomb was exploded at the Nevada Test Site. Thirty-six hours later, an intense rain storm occurred in the city of Troy, New York, 2,300 miles distant from Nevada, and radiation counters at Rensselaer Polytechnic Institute (RPI) began recording atomic fallout three times as high as natural background radiation [1]. Radioactive debris falling onto Troy became an important news event and suddenly the public began to understand that you didn't have to live near the Nevada Test Site to be irradiated by atomic bomb tests.

That event started Barry Commoner thinking in a new direction—a direction that would eventually result in the development of many of the key ideas that now underpin the grass-roots environmental movement.

When Commoner tried to learn more about atomic fallout in 1953, he found that much of the information was secret—classified so by the U.S. government. This made it impossible for academic scientists to examine the data—a violation of the principle of open disclosure that is fundamental to the method of science.

As Commoner expressed it, science can only work when scientists can communicate freely their data and their interpretations of the data. "We need recall," he wrote in *Science* magazine in 1958,

> that the development of a scientific truth is a direct outcome of the degree of communication which normally exists in science. As individuals, scientists are no less fallible than any other reasonably cautious people. What we call a scientific truth emerges from investigators' insistence on free publication of their own observations. This permits the rest of the scientific community to check the data and evaluate the interpretations, so that eventually a commonly held body of facts and ideas comes into being. Any failure to communicate information to the entire scientific community hampers the attainment of a common understanding [2].

The heart of science is open communication, so secrecy—whether imposed by government or by private corporations—is antithetical to science.

Commoner restated many times his view that the scientific method rests squarely on open communication:

> Scientists are, individually, no more truthful than anyone else. Nevertheless, science is a way of getting at the truth, and scientists—most of them—practice their craft in truthful ways. Why? The reason is that science gets at the truth through open discourse. Scientists learn how to practice science truthfully by making their mistakes in public. This permits their colleagues to correct mistaken information and modify faulty conclusions. This is the meaning of open publication of scientific results; it is the essential way in which science approaches the truth [3, p. 7].

The issue of atomic fallout occupied Commoner for a dozen years. While studying it, he derived many of the principles of environmental protection that now form the unspoken basis for grass-roots environmentalism.

MORAL WISDOM RESIDES IN THE CITIZENRY

For example, he clearly established the principle that, in a democracy, scientists have no more right to make decisions than anyone else. Today grass-roots activists might express this as "It's your world. Don't leave it to the experts." Commoner said the same thing forty years ago: decisions with major social consequences must not be left to experts. On the contrary, Commoner said, experts have an obligation to inform the public about the scientific facts and then let the public decide:

> Anyone who attempts to determine whether or not the biological hazards of world-wide fallout can be justified by necessity must somehow weigh a number of human lives against deliberate action to achieve a desired military or political advantage. Such decisions have been made before—for example, by military commanders—but never in the history of humanity has such a judgment involved literally every individual now living and expected for some generations to live on the earth [2].

He went on to ask, who should make such judgments, which require a determination of the value of human life: scientific experts or elected political representatives? [2]

SCIENTISTS HAVE NO SPECIAL COMPETENCE IN MORAL WISDOM

Commoner pointed out that scientists have no special competence in matters of moral judgment. Further, he said, "scientists must take pains to disclaim any special moral wisdom" on the issue of continued above-ground nuclear testing. Scientists should speak on the issue, if they have relevant information to convey, but their expertise does not confer upon them any special capacity to draw moral conclusions from their data. When it comes to balancing citizens' lives against military goals, a scientist is just one more citizen making a moral judgment—his or her scientific expertise has no bearing on the moral equation.

Commoner wrote, "[W]e must not allow this issue [nuclear testing], by default, to rest in the hands of the scientists alone. A question of this gravity cannot be handed over for decision to any group less inclusive than our entire citizenry" [2].

Indeed, it is "self-evident," Commoner argued in 1958, that "the public must be given enough information about the need for testing and the hazards of fallout to permit every citizen to decide for himself whether nuclear tests should go on or be stopped" [2].

CITIZENS CAN INFORM THEMSELVES AND TAKE CHARGE

"Sometimes it is suggested," Commoner wrote,

> that since scientists and engineers have made the bombs, insecticides, and autos, they ought to be responsible for deciding how to deal with the resultant hazards. But this would deprive everyone else of the right of conscience and the political rights of citizenship. This approach would also force us to rely on the moral and political wisdom of scientists and engineers, and there is no evidence that I know of which suggests that they are better endowed in this respect than other people.
>
> There is an alternative, which, though difficult, is feasible. I believe that citizens can continue to rely on their own collective judgment about the issues of environmental conservation—if they take steps to inform themselves [7, p. 180].

ALLIANCE OF SCIENTISTS AND CITIZENS: THE BABY TOOTH SURVEY

Commoner put his ideas into practice: he helped organize scientists and citizens into the St. Louis Committee for Nuclear Information (CNI). They started a newsletter called *Nuclear Information,* which evolved into a magazine with the important (and telling) title, *Scientist and Citizen.*

Commoner, and his fellow scientists at CNI in St. Louis, formed a working alliance with many local citizens. Commoner's work studying atomic fallout had convinced him that fallout represented a biological hazard to humans. However, the U.S. government insisted that fallout was benign. President Eisenhower reflected this official view when, in 1956, he said, "The continuance of the present rate of H-bomb testing[,] by the most sober and responsible scientific judgment . . . does not imperil the health of humanity" [4].

CNI put out a call to the parents of children and began collecting baby teeth and sending them to a lab for analysis of their radioactivity. CNI's goal was to show that strontium-90, one of the main components of fallout from A-bomb testing, was building up in humans. They succeeded. Eight years later, the official U.S. position on atomic fallout had changed completely. In a televised address, in 1964, President Johnson said, "The deadly products of atomic explosions were poisoning our soil and our food and the milk our children drank and the air we all breathe . . . Radioactive poisons were beginning to threaten the safety of people throughout the world. They were a growing menace to the health of every unborn child" [4]. In fact in 1963, President Kennedy had signed an international treaty phasing out above-ground testing of nuclear weapons. It was a triumph of citizen action, working with scientists who helped bring critical facts to light.

Commoner has often acknowledged the important role of an active citizenry:

> Nor is the collaboration between scientist and citizen a one-way street. Citizens have contributed significantly to what scientists now know about fallout. Through the St. Louis Baby Tooth Survey, the children of that city have contributed, as of now, some 150,000 teeth to the cause of scientific knowledge about fallout . . . By such means, and through hard work and financial support many citizens have become partners in the scientific effort to elucidate the fallout problem" [5].

POLLUTION PREVENTION

From this story, we can learn that Commoner pioneered another aspect of modern thinking about the environment. Notice that he did not call for less atomic testing. He called for an end to atomic testing. His training as a biologist convinced him that human intrusions into the global biosphere would have unsuspected consequences:

> . . . [W]henever the biological system exposed to a possibly toxic agent is very large and complex, the probability that any increase in contamination will lead to a new point of attack somewhere in this intricate system cannot be ignored. Finally, the toxic effects of many organic pollutants, like those of radiation, may appear only after a delay of many years. For these reasons, it is prudent to regard any addition of a potentially toxic substance to the biosphere as capable of producing a total biological effect which is roughly proportional to its concentration in the biosphere [5].

In this passage, Commoner is saying there is probably no threshold for pollution effects—any amount of pollution can be expected to cause some damage. Thus, the only way to prevent environmental damage from toxic pollution would be to exclude pollution from the environment completely. Today we call it pollution prevention. Barry Commoner argued for it, and provided the rationale for it, nearly forty years ago.

In 1964, Commoner also warned: "Like radiation, many of the new synthetic substances act on basic biochemical processes that occur in some form in all living things. Hence, we must anticipate some effects on all forms of life" [5, p. 6].

BURDEN OF PROOF

The preceding two quotations reveal another aspect of Commoner's thinking in the late 1950s and early 1960s.

At that time, most scientists believed that biological systems had a fixed capacity to assimilate pollutants, to absorb insults without harm. If this view were correct, then scientists would merely need to discover the "assimilative capacity" of a biological system (a forest or wetland or bird or human) and then government officials could make regulations that would prevent pollution from exceeding the assimilative capacity. Polluters would not need to show that their pollutants were harmless—the regulatory system would deem them harmless as a matter of science and of law.

Commoner articulated a different view. If any amount of pollution would be expected to cause some damage, and if every form of life would be subject to damage, then the proper level of pollution would not be discoverable by science. It would require a moral judgment by citizens. This view would later be used by grass-roots activists to provide the foundation for shifting the burden of proof of safety onto the polluters.

Here is how it works today: If we expect any amount of pollution to cause some damage, then the burden of proof must logically shift to the polluter, who must convince an informed citizenry that his or her pollution will (unexpectedly) cause no damage or will cause only an "acceptable" amount of damage. Naturally, citizens insist that "acceptable" pollution levels are not a scientific matter but a political and moral judgment.

RIGHT-TO-KNOW

In 1964, Commoner called for "a registry of the events which disseminate pollutants into the environment"—which today we might call a "right-to-know database."

He wrote,

> Thus, if we are to balance the risks and benefits involved in the Mississippi River pollution problem, experience with fallout tells us how this can be done: by frequent and detailed monitoring of the pollutant, and by registry of the

events which disseminate pollutants into the biosphere . . . Unless there is a registry of the amount of insecticides released into the river by agricultural and by industrial operations, we cannot know what benefits are to be balanced against the risks [5, p. 6].

And he said again,

The lessons of this [atomic testing] experience are self-evident. Massive and pervasive contamination such as that due to fallout can indeed be understood and controversies as to its effects resolved, if we are aware of the sources and report the dissemination of the contaminants by a system of detailed monitoring [5, p. 4].

Congress finally got around to creating the "right-to-know database"—formally called the Toxics Release Inventory—in 1985.

THE PRECAUTIONARY PRINCIPLE

In 1964, Commoner foreshadowed one of the key ideas of modern environmentalism—the precautionary principle. He said, "Until we can balance known risks against specific benefits, no meaningful decision as to the action required by these pollutants is possible. But even in the absence of such a decision, the rule of prudence—which is demanded by the unknown long-term hazard—requires that extreme caution be exercised in continued use of these agents."

And he said, ". . . I would suggest that our knowledge of these basic problems is still so rudimentary that great prudence ought to be the guide in the introduction of any new organic substance into the water supply" [5, p. 7].

In 1964, this was a new idea indeed. However, today it is quite mainstream. A policy of prudence was adopted in 1992 at the Rio de Janeiro UN Conference on Environment and Development (UNCED). Principle 15 of the Rio Declaration endorses the precautionary approach (and pollution prevention) in the following terms:

In order to protect the environment, the precautionary approach shall be widely applied by States according to their capabilities. Where there are threats of serious or irreversible damage, lack of full scientific certainty shall not be used as a reason for postponing cost effective measures to prevent environmental degradation.

The crux of the precautionary principle is: action to prevent serious or irreversible damage should not be delayed until the scientific evidence is clear because by that time it may well be too late.

Commoner advocated it thirty years before the world community embraced it.

ENVIRONMENTAL IMPACT ASSESSMENT

In 1964, Commoner called for environmental impact assessment. He did not use those words, but he described the concept:

> Most important, we must understand that our present difficulties are due to the large-scale dissemination of substances that have not been subjected to adequate biological analyses on the scale in which they are used. And such an understanding ought to be reflected in a resolve to require from now on that newly synthesized compounds be tested in the natural environment before they are committed to large scale economic investment and use [5, pp. 6-7].

And in 1965 he wrote,

> The scale of new scientific and technological operations places new strains on the process of scientific discourse. What are required are means for bringing these new activities into more effective contact with the system of discourse that science has already established. Large-scale experiments and technological operations that might lead to violations of the principles of disciplined experimentation, or which might interfere with controls necessary for other research, ought to be exposed to open consideration by the scientific community before they are undertaken [6, pp. 195-196].

In late 1969 Congress passed the National Environmental Policy Act (NEPA). One of the key features of NEPA was the requirement that the federal government must assess the environmental impacts of any significant expenditure of its funds. The concept of assessing the impacts of large-scale technologies has not yet been carried over into the private sector of the economy, but several states have adopted it, and it would seem only a matter of time before the private sector is required to assess the impacts of large-scale technical innovations before they are deployed.

RISK ASSESSMENT IS POLITICAL

Discussing the role of scientists in controversies involving nuclear fallout, nuclear war, or "environmental contamination in general," a committee chaired by Barry Commoner wrote in 1965,

> In a number of instances, individual scientists, independent scientific committees, and scientific advisory groups to the government have stated that a particular hazard is "negligible," or "acceptable" or "unacceptable"—without making it clear that the conclusion is not a scientific conclusion, but a social judgment. [Substitute the word "risk" for "hazard" in that last sentence and notice the modern ring that it takes on.—P.M.] Nevertheless, it is natural that the public should assume that such pronouncements are scientific

conclusions. Since such conclusions, put forward by individual scientists, or by groups of scientists, are often contradictory, a question which commonly arises among the public is, "How do we know which scientists are telling the truth [6]?

The preceding paragraph sums up very well the modern dilemma with "risk assessment," which is this: the judgment of acceptable risks, or hazards, is a political and moral matter. Yet scientists are hired by industry and government to assess the risks of various technological intrusions into the environment and the public may therefore mistakenly come to believe that the judgment of acceptable risks is a scientific matter.

Commoner identified this problem in the setting of radiation standards for the nation:

> The growing interaction between science and public policy requires considerable attention to the problem of distinguishing scientific problems from those issues which ought to be decided by social processes. An example is the tendency to confuse scientific evaluation with social judgment in the matter of radiation standards. Here, a scientific body, the federal radiation council, is engaged in setting standards of acceptability which are basically social judgments regarding the balance between the hazards and benefits of nuclear operations. These judgments are, or ought to be, wholly vulnerable to political debate, but their appearance in the guise of a scientific decision may shield them from such scrutiny [6, p. 194].

Thus in 1965 Commoner and his colleagues warned us that risk assessments are political in nature, not merely scientific, and that many scientists overstep the bounds of scientific legitimacy and try to impose their (or their employer's) political decisions and views upon the public, using science as a screen.

These ideas about risk assessment seem entirely contemporary and universal because they are deeply held today (based on experience) by grass-roots environmentalists everywhere. But really these ideas only sound contemporary and universal. They are at least thirty-five years old, and they came to us through the clear thinking and hard work of one man of extraordinary vision—Barry Commoner.

REFERENCES

1. H. M. Clark, The Occurrence of Unusually High-Level Radioactive Rainout in the Area of Troy, N.Y., *Science, 119*, pp. 619-622, May 7, 1954.
2. B. Commoner, The Fallout Problem, *Science, 127*:3305, pp. 1023-1026, May 2, 1958.
3. B. Commoner, *Toward a Humane Science,* Reed College Sallyport, August 1970.
4. Quoted in Barry Commoner, The Myth of Omnipotence, *Environment, 11*:2, pp. 8-13, 26-28, March 1969.

5. B. Commoner, Fallout and Water Pollution—Parallel Cases, *Scientist and Citizen*, 7:2, pp. 2-7, December 1964.
6. AAAS Committee on Science in the Promotion of Human Welfare, The Integrity of Science, *American Scientist, 53*, pp. 174-198, 1965.
7. B. Commoner, Duty of Science in the Ecological Crisis, *Scientist and Citizen*, pp. 173-182, October 1967.

CHAPTER 2

The Day Before Yesterday: The Committees for Nuclear and Environmental Information

Virginia Warner Brodine

Barry Commoner's leadership in the formation and early years of the Committee for Nuclear Information is described. The Committee's role as a pioneer in providing the public with information on nuclear questions, then the prime environmental issue requiring political action, is outlined. When it changed its name to the Committee for Environmental Information and broadened its scope, the focus continued to be on those environmental issues requiring political decisions. Although both Committees limited themselves to scientific information and did not advocate particular political solutions, they became embroiled in controversies, some of them significant for breaking through barriers of government silence and corporate misinformation.

The modern environmental movement began only yesterday, its birth usually dated in the early 1960s with the publication of Rachel Carson's *Silent Spring*. The task of this movement is to understand the effects of resource use and production processes on people and nature and make that understanding part of economic and political decisions.

The Committee for Nuclear Information (CNI) was one of the first organizations devoted to that task. Its story takes us back into the '50s—the day before yesterday.

During our first five years, the focus was solely on nuclear energy. The threat of nuclear war was at that time the greatest threat to both people and the natural environment, while preparations for nuclear war presented their own dangers, as did peaceful uses of nuclear energy.

The subject matter broadened in 1963 and the name of the organization changed to Committee for Environmental Information (CEI). Since we were already in business, we were in a good position to provide the new and growing movement with environmental information.

All of us who participated in CNI/CEI can be proud of its historic role but it was Barry Commoner who invented it, and who was its preeminent leader and spokesman. He actively participated in its work, wrote for its publication, guided its direction, and was primarily responsible for obtaining the grants that made its work possible. His ability to convince and to inspire drew people to CNI. It was one of the reasons that not only Commoner himself, but many others did the important work that made CNI/CEI's twenty-year history a significant contribution.

HOW CNI STARTED

Now, a brief sketch of how CNI started and how it operated, from my personal point of view.

My participation began on a Sunday evening in the fall of 1956. As Barry Commoner tells us in *The Closing Circle*, he was at that time becoming increasingly concerned about fallout from atmospheric nuclear testing [1]. As a personal friend, I had been learning about it from him.

On the evening in question, he came up with what I thought was a great idea.

"It wasn't so many years ago that St. Louis women led a successful campaign for pure milk," he said. "What if there were a similar campaign, this time to get the milk tested for strontium-90?"

I got a positive response to the idea the next day from two of my colleagues on the regional staff of the International Ladies' Garment Workers Union. Edna Gellhorn, one of the women who had led the pure milk campaign, soon joined us and brought others with her. Some women wanted to participate because they had heard about fallout from physicist John Fowler and pathologist Walter Bauer, two of Commoner's Washington University colleagues who had joined him in taking fallout information to citizen groups.

A letter requesting that our milk be tested for strontium-90 went off to the U.S. Public Health Service and to the St. Louis City Health Department with the signatures of eighteen women. This was the first step in our efforts to spread our concern.

Then came another great Commoner idea: The citizens and the scientists should get together and establish an organization to inform the public—in St. Louis and elsewhere—of the scientific facts about fallout and other nuclear issues. Preliminary work developed a large and diverse list of sponsors and in April 1958 the Greater St. Louis Citizens Committee for Nuclear Information was founded.

It is not possible to mention everyone who made that happen and who got CNI established and functioning in the next five years, but the organizational skills and commitment of two women stand out: Gloria Gordon and Judy Baumgarten.

A POLITICAL PURPOSE

CNI was not a grass-roots organization in the sense of local membership activity and control. It did not take political positions, but it had a definite political purpose: Some significant political decisions needed to be made about nuclear energy. They should not be made in ignorance nor by unchallenged power, but by an informed citizenry.

CNI was unique in two ways. First, it organized scientists into a Technical Division for a specific and focused social purpose. Second, although it was concerned with science, it drew its membership, officers, and board from other citizens as well as scientists.

The job the Technical Division took on required a big commitment of time and a lot of courage. Open and rational discussion of nuclear issues was extremely difficult in the political atmosphere of the '50s. To challenge nuclear weapons in any way was to lay oneself open to accusations of being unpatriotic and playing into the hands of our designated enemy, the Soviet Union. Euphoria about the technological magic of nuclear energy rendered non-military uses almost as sacrosanct.

The biological effects of the use of nuclear energy were just beginning to be understood by scientists. The Atomic Energy Commission (AEC), the government's overseer of all things nuclear, tended to publicize only such information as would reflect positively on its projects. The country as a whole remained in the dark.

None of the CNI scientists was a nuclear physicist or radiation biologist. Thus they had to buck the cult of the expert as well as the power and prestige of the government—and government prestige was high in those pre-Watergate days.

A Speakers Bureau drew on the services of twenty scientists who spoke to fifty groups of citizens in the first year [2]. A mimeographed information bulletin was issued sporadically. In the following years, it grew into a monthly magazine. Its content was the responsibility of the Technical Division, a heavy responsibility in the early years when the work was all voluntary, somewhat less burdensome beginning in 1962 when professional staff was hired. The magazine was called *Nuclear Information* until August 1964 when it became *Scientist and Citizen*. In January 1969, its name changed once more to *Environment*.

Members of the division wrote for the magazine and reviewed all articles submitted for publication. When a major statement was required, for example for Congressional testimony, it might be prepared by one or more individuals but the Technical Division discussed it collectively and it came out with the imprimatur of the division as a whole.

COMMONER'S CONFIDENCE

What carried not only the Technical Division but the whole organization along more than anything else was Barry Commoner's unwavering confidence in the importance of the information and the ability of the public to understand and use it.

The citizens who served as officers and board members or who were staff members or volunteers in the office knew they were involved in a vitally important effort. Although our work was essentially that of assisting the scientists, the relationship was close in the early years and the controversies in which CNI became involved were exciting.

As editor of the magazine from 1962 to 1969, I worked with the Technical Division, especially with Barry Commoner and with the three scientists who successively chaired the division: Eric Reiss, Malcolm Peterson, and Kurt Hohenemser. It was a great learning experience.

When CNI was only a few months old, it embarked on a new project which tied it more firmly to the community and could hardly have been done by scientists alone. This was the Baby Tooth Survey, a study of children's teeth for strontium-90.

Initiated by a CNI committee of which Commoner was a member, it met with skepticism at first. Could enough ordinary people understand and be willing to participate in a scientific research project? Under the leadership of Dr. Louise Reiss, hard work by a citizens' committee, a respectful approach to the community and some very creative publicity ideas succeeded in gaining the support of community organizations and the St. Louis school system [3]. The Washington University Dental School obtained a grant for the study but the collection of the teeth begun by CNI continued under the direction of Yvonne Logan to be part of CNI's operation. Tooth collections and comparative studies were initiated in other cities. The importance of the data the study produced was matched by its significance in educating people to the meaning of atmospheric testing [4]. They could see as no mere words could show them that radioactive fallout was getting into their children's bodies.

A few of the CNI scientists, Barry Commoner in particular, knew how to write for the public but most welcomed editorial help. A few CNI members gave them this help in the early years but from 1962 on that aspect of bridging the scientist-citizen gap was taken over by the magazine staff. In 1965-66, we had the help of a Readers' Advisory Committee which critiqued the magazine's material for readability [5].

One way to evaluate CNI/CEI's work is to ask the questions: Did our information get into the hands of the people who needed it? Did they use it politically to effect any differences in policy?

PUBLIC PERCEPTION CHANGED

A change certainly took place in the public perception of the dangers of nuclear energy in the twenty years of CNI/CEI's life. Our work can realistically be considered one factor.

To get down to specifics, here are four examples of our information in the nuclear field.

1. Fallout from atmospheric nuclear testing was an important focus of CNI until the Partial Test Ban Treaty in 1963. Testimony before the relevant Congressional committee, and the controversy this sparked with the AEC [6] helped the information reach a national audience. This and other controversies attracted media attention which spread our information further.

The treaty was a contentious issue and even after it was signed by President Kennedy, it had strong opposition. However, it had wide *and* informed public support and was ratified overwhelmingly.

2. Fallout shelters and other civil defense measures were part of preparations for nuclear war throughout this period. In a series of eleven issues on nuclear war and civil defense, two of them by Commoner, CNI put important information on the weapons themselves and on the problems of economic and social recovery from nuclear war into the hands of concerned citizen activists. CNI showed that any civil defense program intended to permit the nation to survive a nuclear attack would be a monumental task with an uncertain end [7].

A National Academy of Sciences–National Research Council summer study group on civil defense took place in 1963. Called "Project Harbor," the study was done in response to a request from the Department of Defense and the AEC, and was headed by Nobel laureate Eugene Wigner.

In spite of many CNI requests for the results of the study, nothing was forthcoming for a year-and-a-half. Finally, a Summary Report was received. As CNI later pointed out, this document discussed the tactics of persuading the public to accept civil defense measures rather than objectively presenting the facts and letting the public and its non-scientific representatives decide on the basis of the facts [8].

REQUEST REFUSED

CNI asked for the full report, but was refused on the grounds that it was "available only to those federal agencies concerned." After repeated insistence that it be publicly available, CNI finally got a copy on loan from the Army Library.

A lively discussion in three issues of our magazine followed [9]. The Harbor Summary was published and criticized by CNI for failure to consider some important aspects of recovery and for over-optimistic assessments of the ability of a civil defense program to protect. Wigner and other Project Harbor participants responded. One paper from the full report that had not been reflected in the summary was published. A number of other scientists, including some highly regarded members of the National Academy of Sciences, added their own comments critical of the Harbor Report.

There was never a groundswell in favor of civil defense. Although government agencies put billions of dollars into civil defense in the following years, the major blast shelter program advocated in Project Harbor was never instituted.

3. Project Plowshare was a favorite "peaceful use" of the AEC. It proposed the use of nuclear explosives to blast a harbor, a canal, or tunnel for a rail and highway route. Some states planned similar projects of their own. Mention such an idea today and you will be met with total disbelief that it could ever have been considered seriously. Yet from 1960 to 1969 grandiose plans were put forward and some tests were carried out. Separate projects within Plowshare were given quaint vehicular titles, such as Project Sedan, Project Carryall, Project Gasbuggy, and Project Scooter.

Plowshare drew the attention of CNI in 1960 [10]. In 1961, an issue on Project Chariot, a proposal to dig a harbor in Alaska, was prepared in cooperation with Alaska scientists and the Alaska Conservation Society [11]. It was attacked in an article in *Science* to which CNI replied in the same journal.

The project was dropped when the international moratorium on atmospheric testing ended and AEC attention turned to weapons testing.

In 1969, a joint AEC-Mining Company harbor in Australia was proposed. Australian scientists, using CNI materials, demanded extensive preliminary ecological and oceanographic studies and an independent committee to evaluate the safety of the project when the studies had been completed. Project Plowshare had plowed itself into the water and never emerged [12].

OPPOSING A REACTOR

4. A citizen group opposing the building of a nuclear reactor by Pacific Gas and Electric Co. (PG&E) at Bodega Bay in California called CEI for help. They could not find a scientist in the area willing to discuss its possible hazards. A speaker from CEI responded to the call. Subsequently, an article in *Nuclear Information* focused on the problem of the reactor's siting, close to the San Andreas Fault [13]. It was widely reported in the California press. PG&E, which already had a $4-million dollar hole on Bodega Head called it "clever and dangerous sabotage," [14] and sent their public relations man to St. Louis. He visited the CNI office and the *St. Louis Post-Dispatch* which had called the article "a service to the nation."

The AEC's regulatory branch sent PG&E a letter raising many of the questions about the reactor's safety previously raised by CNI. A few months later, the citizens' group celebrated a victory: PG&E announced that it was dropping plans for the Bodega reactor [15].

Important as nuclear energy was, the organization cannot be evaluated on that alone. It is much more difficult, however, to assess CEI's effect in the wider environmental area where the political decisions were apt to be more complex and difficult than whether to halt atmospheric nuclear testing or to stop the nuclear blasting of a harbor.

Throughout the '60s, the new subject matter had given the organization new vigor and brought new scientists to the Technical Division which developed committees on air pollution, water pollution, and pesticides.

At the same time, the magazine was becoming more truly national, with contributing scientists from all over the country. In other cities, individual scientists and groups had begun, not long after CNI started, to undertake similar tasks. A conference in New York brought them together in 1963 and established the Scientists Institute for Public Information (SIPI). All the science information groups used our magazine and in 1967 it became SIPI's official publication.

NEW SUBJECTS

New controversies followed the new environmental subjects.

A two-year Interstate Air Pollution Study in the St. Louis-East St. Louis metropolitan area was carried out by the federal Public Health Service and local health officials in 1963-64. While it was underway, CEI speakers were active in educating the public about air pollution and were regarded as valuable allies by the agencies involved. When the report of the study was released, it was carefully evaluated by the CNI Air Pollution Committee [16].

Hearings on proposed air standards brought "experts" to St. Louis to testify on behalf of the industries most affected. One of them questioned the correlation of air pollution with the increased incidence of respiratory ailments in urban areas. According to him, the increase might just as well be correlated with increases in tv dinners, telephones, and miniskirts [17].

The Air Pollution Committee countered this expertise but also criticized the limitations of the report and its proposed standards, which were not designed to clean up the air, but merely to keep it from getting dirtier. The Public Health Service was outraged. For some reason, it focused on me as editor and sent a representative to St. Louis, to—as he said—"straighten that woman out."

Our discussion took place in the Medical School Library. The straightening out process was competently handled by Malcolm Peterson, Technical Division Chair, and members of the Air Pollution Committee, chaired by Robert Karsh. "That woman" enjoyed the occasion.

A THREAT TO SUE

Pesticide issues brought us into conflict with chemical corporations. A draft article on the Shell Company's No-Pest Strip produced the most extreme reaction: a threat to sue if it were published.

The board had a serious discussion of the expensive court case that might ensue if the threat were carried out and the fact that if CEI lost, the board would be individually and collectively responsible for paying a fine. The vote to publish was unanimous.

The article was revised and strengthened in the light of further information. It was, of course, carefully reviewed by the scientists and came out under the imprimatur of the committee as a whole [18]. Shell did not sue, but the pest strip is still on the market.[1]

A second article in the same issue detailed the efforts to obtain the scientific studies used by Shell to justify its claims of harmlessness. The regulatory process that permitted the manufacture and sale of the No-Pest Strip with—or without—labels limiting the way it should be used was also described [19].

These articles make interesting reading today in the light of current concern over the regulation of toxic chemicals [20]. In fact, it is remarkable how timely are many of *Environment's* articles throughout the period when it was edited by Sheldon Novick, Julian McCaull, and Kevin Shea (1969-77), and even from its earlier years.

It was not that CEI was ahead of its time but that a publication limited to information, important as that is, can do only one part of the job. There is still a gap between the political and economic changes imperatively called for by the information and the ability of citizens to make those changes.

Barry Commoner, with his science and social activism, influenced many people in CNI/CEI, as he has done since. Perhaps it should be considered another of his contributions to the environmental movement that some of us are trying to find ways to close that gap.

REFERENCES

1. B. Commoner, *The Closing Circle*, Alfred P. Knopf, Inc., New York, pp. 45-51, 1971.
2. *Information From the Greater St. Louis Citizens Committee for Nuclear Information, 1*, December 4, 1958.
3. Y. Logan, The Story of the Baby Tooth Survey, *Scientist and Citizen, 6*, pp. 38-39, September-October 1964.
4. W. Krasner, Baby Tooth Survey—First Results, *Nuclear Information, 4*, pp. 1-6, January 1962. Based on an article by L. Z. Reiss in *Science, 24*, November 1961, H. T. Blumenthal, Strontium-90 in Children, *Scientist and Citizen, 6*, pp. 3-7, September-October 1964. Based in part on an article by H. L. Rosenthal, S. Austin, S. O'Neill, K. Tahouchi, J. Bird, and J. E. Gilster, Incorporation of Fallout Sr-90 in Deciduous

---

[1] In the course of preparing this talk, I checked with the EPA Pesticide Division and learned that Shell Chemical Co. has forty products registered which include percentages of DDVP ranging from 20 to 44 1/2. The basic material for resin pest strips has 20 percent. Shell informed me that it no longer markets the No-Pest Strip, having sold it to the Loveland Chemical Co. A call to Loveland elicited the information that it markets the pest strips under the name of Prozap Insect Guard for use in homes, cabins, garages, motels, and so forth. EPA regulations require labels stating that it is not for use in kitchens, restaurants, or food processing plants, or in hospitals, clinics, or patient rooms anywhere.

Incisors and Foetal Bone, *Nature, 203,* p. 615, 1964, and Baby Tooth Survey, A Comparative Study, *Scientist and Citizen, 8,* pp. 16-17, May 1966.
5. M. Cunningham, *The Citizen's View of Scientist and Citizen, 8,* pp. 7-18, May 1966.
6. Technical Division, Committee for Nuclear Information, Local Fallout: Hazard from Nevada Tests, *Nuclear Information, 5,* pp. 1-12, August 1963, and Hazard from Nevada Tests, a reply to the Atomic Energy Commission, *6,* pp. 1-2, November 1963, and G. M. Dunning, Discussion of Iodine 131 in Fallout with *CNI Comments,* in the same issue, pp. 3-13.
7. B. Commoner, Civil Defense—The Citizen's Choice, *Nuclear Information, 6,* pp. 1-17, June-July 1964.
8. Scientific Advisory Board of *Scientist and Citizen,* Concluding the Project Harbor discussion, *Scientist and Citizen, 8,* p. 31, February-March 1966.
9. *Scientist and Citizen, 7,* May-June and August 1965, and *8,* p. 8, February-March 1966.
10. C. Hohenemser, Blasting a Harbor, and What is Plowshare, with Barry Commoner, *Nuclear Information, 2,* pp. 1-4, June 1960, and More on Plowshare, *2,* pp. 1, 3-4, July 1960.
11. Project Chariot, 3 (1961): June including B. Commoner, M. W. Friedlander, C. Hohenemser, and E. Reiss, *The Known and the Unknown—The Nature of Nuclear Decisions,* M. W. Friedlander, *Predictions of Fallout from Project Chariot,* B. Commoner, *Biological Risks from Project Chariot,* D. C. Foote, *Cape Thompson: The Place, The People,* and W. O. Pruitt, Jr., *The Animals.*
12. V. Brodine, Six Questions for Australians, *Environment, 11,* pp. 16-19, April 1969, and *Unsnug Harbor,* pp. 34-36, May 1969.
13. L. Mattison, A Quake at Bodega, *Nuclear Information, 6,* pp. 1-12, April 1964.
14. More on Bodega, *Nuclear Information, 6,* pp. 11-13, August 1964.
15. *New York Times,* October 27 and 30, noted in *Scientist and Citizen, 7,* inside front cover, November 1964.
16. Air Pollution Committee, CNI Technical Division, Air Quality in the St. Louis Area, *Scientist and Citizen, 8,* pp. 9-10, January 1966, and The Air St. Louisans Breathe, pp. 6-15, May 1966; including R. Karsh, The St. Louis Sulfation Story, R. B. Ferguson, Automobiles and Air Pollution, and P. Gaspar, Industrial Process Emissions.
17. J. Shulman, New Standards in the Making, *Scientist and Citizen, 9,* pp. 16-17, 1967.
18. Committee for Environmental Information, The Price of Convenience, *Environment,* pp. 2-15, 44-48, 1970.
19. S. Novick, The Burden of Proof, *Environment,* pp. 16-29, 1970.
20. *Rachel's Health & Environment Weekly,* "On Regulation," #538, March 20, 1997.

CHAPTER 3
Crossing Paths: Science and the Working Class

Tony Mazzocchi

My first meeting with Barry Commoner developed out of my interest in nuclear weapons testing, an issue that had engrossed Commoner and the Committee for Nuclear Information (CNI) for some time. Then, our union local worked collaboratively with the scientists on the Baby Tooth Survey which showed an association between nuclear fallout and the level of strontium-90 in children's teeth. Linking the CNI scientists and the working class in this way produced results and Commoner and his colleagues made themselves available to respond to corporate scientists on issues such as toxic workplaces exposures. We were successful in raising occupational health and safety questions because we had credible scientists on our side. Through the association with Barry, we changed the face of the workplace and thousands of people are alive today because we helped pass the federal Occupational Safety and Health Act. A whole cadre of health and safety specialists came out of those efforts.

It is a privilege to be here today and see some very old friends and realize that life is full of intersections and some of them have profound impacts on the way things happen. It is a special privilege to talk at Barry Commoner's eightieth birthday party because when my path crossed with Barry's and CNI's, it changed me and, by virtue of changing me, it affected a lot of other lives.

It all started back in 1956. I was a union president of my local out on Long Island. Adlai Stevenson was running for president and my office was right in town, which was rather conservative at the time. It was called Roslyn, Long Island, and our office was right dead set in the middle of town by a clock tower. Our local endorsed Adlai Stevenson for president and we put this big sign outside our office and people tracked in and talked to us. It turned out that that was the first time a sign for a Democratic candidate for anything had ever been put up in that town; it had a tourist reputation for being a Republican stronghold. It has certainly changed since then. We endorsed Stevenson, but we really didn't know how to carry out a campaign.

IDEA FOR A MEETING

I had been reading that Stevenson had come out against nuclear weapons testing and I knew nothing about the subject matter but it sounded like an interesting topic and maybe our local union could sponsor a meeting about it. As a result, I was told about a radio correspondent who worked for CBS; I remember her name was Judith Levitan. She came down to the office and she said she knew this nuclear physicist who had just come back from the Bikini test, and that sounded interesting. So I said, well, if we have a meeting and we get this scientist, he could tell us all about nuclear weapons testing and fallout. It turned out that he brought along two other people, a pathologist and, I think, a geneticist who had participated in the test. We leafletized the area. We figured we'd get 100 people out to a meeting in the high school. Eight hundred people showed up, and these three scientists. By the way, they are the first scientists in the flesh I have ever met in my life. I'm thirty years old in 1956, and I meet, for the first time, a scientist, a real live scientist.

They get up on the platform at this meeting with 800 people and they get scared to death. It was one of these deals where, you know, testing, on the one hand, might be bad for you, but on the other hand, it might not be so bad. Everyone walked out of there frustrated as hell. If you think it was bad for anybody else talking about nuclear weapons testing in 1956, to come out of a union like mine where a substantial part of our membership was making bombs, at the height of bomb production in 1956. . . . But after the meeting, I became interested in the subject and said, boy, I guess there are a number of scientists out there who won't be as scared as these guys are and would be willing to talk.

I got introduced to various scientists around the city of New York. The first scientist said, "I can't talk about it, but I'll tell you somebody who can." And I trudged all over the city of New York, even met a Nobel Prize winner, first one I ever met. And he said, "Well, you know, there are both sides to this question."

And I said, "Yes, but just tell us what the information is. I think people ought to know about it."

He was scared and then someone else told me he had a grant from the Atomic Energy Commission (AEC) and I began to understand that no one was going to talk about this subject matter and I gave up.

Stevenson proceeded to lose the election but he called back to us on the island and said that ours was the only meeting during the entire 1956 campaign devoted exclusively to nuclear weapons testing, so we put some small comfort in that.

A FOCUS ON TESTING

However, after the session was over, it was early 1957 and a Quaker by the name of Stuart Meecham contacted me and said, "I hear you're interested in the issue of nuclear weapons testing."

And I said, "Yes, but nobody else is interested."

He said, "Well, we're having a meeting in New York at the Woodrow Wilson Foundation."

And that was quite a couple of months. I had never met a scientist, I met a real scientist, and then I go to this meeting at Woodrow Wilson and there's the most influential members of the Protestant clergy—no Catholics, no Catholic priests and no Jews or rabbis—and here we are in the city of New York. This group was called together to talk about nuclear weapons testing. They were the only group, I guess, that wasn't afraid to talk about it. And we formed a group called the New York Committee Against Nuclear Weapons Testing. I became the treasurer. As a matter of fact, the first contribution to that committee came from my local union. This group evolved into the Committee for Sane Nuclear Policy and, eventually, "SANE."

As a result of this association during the campaign with nuclear weapons testing, I began to understand that there were scientists out there who were willing to speak out on this subject. There were many. And they had never met members of the working class. There were two distinct universes. Then, I hear about this group in St. Louis—The Committee for Nuclear Information (CNI), and I reached out and they, indeed, were prepared to assist in another matter that was arising in trade union consciousness. That was the question of what's happening to us as workers on the job.

The first thought that came to my head was there were no scientists out there willing to talk to us about the association between toxic exposures and subsequent disease that we were observing, and we were looking and called CNI and met many of the people—Virginia Brodine, Barry Commoner and others, names that I now recall rapidly coming back to me. They provided names of scientists, and I remember the first couple were two young men who were Rene Dubos' students, Glenn Paulson and Max Snoderly, who were members of the committee, and Al Nadler was a physician. We formed a dog-and-pony show and tramped all through North America. And when people described their workplace conditions, I had these credible scientists who could respond. That marriage had an incredible effect because later on many medical and industrial hygiene students and others concerned with public health worked with our union. It was the original association with CNI that allowed that to happen. That was the first crossing of members of the science community's path and the path of a working class organization. We began to work in a methodical manner to expose any number of these problems.

OFFERING BABY'S TEETH

But I guess our golden moment working with CNI was, in fact, the Baby Tooth Survey. Eighty percent of the membership of my local union are women, and when we heard about the Baby Tooth Survey, it was a way in which we could link our own members into a greater national question of nuclear fallout. I had been talking about that because I was a member at that time of "SANE" and my local was a sponsor. But this was a tangible way of associating a subject that most people thought was too complicated for them to understand: the question of strontium-90 being taken up by humans and lodging in their bones. Here was a group that came out with a method to demonstrate that there was an association. Our members contributed thousands of teeth. We had grandkids, and every morning someone would come into the local with a little package; you'd write down the age of the child, and so forth. And we would ship those teeth out. It was an incredible educational program because people had a direct relationship to it. It was their kids' teeth, and to think that these teeth had strontium-90 in them. The survey allowed me a political base to talk about this question to the union which was involved in developing nuclear weapons. Without that, I think I would not have survived politically in the union.

So Barry and CNI were an incredible source of information. Not only a source of information, but a supportive place we could reach into and bring out credible scientists to match the industry scientists on all subjects, whether it was toxic exposure in the workplace or a question about nuclear weapons testing.

I recall very well in 1972 when we went to Stockholm to a world environmental conference. I was part of the Scientists Institute for Public Information (SIPI) team that went over. Barry, of course, insisted that we include trade union representatives and I was *the* trade union representative. And we talked about occupational health as a major environmental issue. That was an incredible experience because we thought the Europeans were ahead of us on this matter. And here I was in Sweden talking in union halls and finding out they were even behind us. We just assumed they knew more than we did. And I'm describing what it was like to work in an American workplace and these Swedes came up to me after each meeting and would say,

> "Hey, you've just described the way it is where I work!"
> And I said, "Well I thought you all had laws that dealt with this subject matter."
> They said, "No, no, we have none of that."

That was at that time. They rapidly made a hell of a lot of progress. But it grew out of this meeting in Stockholm where we had this wild group. I think the United Nations allowed this alternative conference to take place. I think we were up in the theater, Ingmar Bergman's Theater Group where we met, and I remember Barry debating Paul Erlich on population and we all got a chance to tell the story from a grass-roots sampling.

But this conference—and Barry organized our participation—had an incredible impact.

EXPOSURES AT WORK

CNI, which ultimately ended up as the SIPI, was the only group that reached out for representatives of the working class. That association was unique, but it allowed us to reach out to a broader segment of the community, and I think we were successful in raising in the minds of the public the question of occupational health and safety because we had a reservoir of credible people who could back it up using their scientific expertise.

A critical moment in the struggle to bring occupational health and safety to public consciousness was in 1973 when our union conducted the first strike around occupational health and safety. We negotiated a major agreement with the oil industry that recognized that workers had a right to say something about that which affected their lives on the job. It was an incredible step and agreement.

Most of the industry, except Shell, agreed to our contract proposals. Shell decided it was going to draw the line and fight. Its position was that health and safety was the prerogative of the company and the potential victim, the worker, really had no say in what affected him or her. They thought this was a principle fight that they had to engage in. We had to call out the entire Shell work force that we represented on a strike that lasted about five months. It was a very difficult strike because we were fighting over a question that people didn't quite fully understand and even the workers themselves were only beginning to understand the association between toxic exposure and subsequent ill health effects and death. That strike dragged on and Barry's group provided us a list of scientists to advise us.

We ended up with two Nobel Prize winners on the committee, along with a lot of other people whose names you would recognize. In fact, Dr. Eula Bingham was our first chair who went on to be probably the best Assistant Secretary of Labor in heading up the Occupational Safety and Health Administration (OSHA) that we had. It was a group of scientists who said, many of them, that they didn't quite know much about trade unions but they understood the relationship between toxic exposure and worker health. As a result of the committee, we had a large press conference and got a front page article in *The Washington Post* and an editorial in *The New York Times*. That was what really brought Shell to their knees. They weren't worried about us as a union. What they were worried about was the publicity and articulation by members of the science community who were tapping public consciousness.

CHANGING THE WORKPLACE

Without that association with Barry, we would not have changed the face of the workplace. Not that it's changed the way we want it to be changed, but I think many thousands of people are alive today because we intervened in the late 1960s and '70s and got a law (the OSH Act). A whole cadre of health and safety specialists came out of that effort and it had a tremendous impact on people's lives. I can say that people are probably alive today who wouldn't be alive because many of those with whom I worked are now dead as a result of workplace exposures.

So on behalf of those untold thousands who are alive, the fact that our paths crossed had an incredible, lasting impact. We continue to work with the science community. In fact, the journal, *New Solutions,* that I started when I was Secretary-Treasurer of our union, grew out of the consciousness that I developed in this association. I knew that we needed one publication that linked science and the working class community, and that *Journal* is the result. I associate it all with that moment in time in the late 1950s and early 1960s when this working class organization crossed paths with Barry Commoner and CNI and all those folks who assisted us.

CHAPTER 4

Real Junk Science: The Corruption of Science by Corporate Money

Ralph Nader

Barry Commoner's eightieth birthday is an occasion to look at the phenomenon of junk corporate science, which must be viewed in the context of the power game. Bought-and-paid-for scientists are instruments of corporate power used in the pursuit of profits to ward off the mobilization of the citizenry and the application of regulatory health and safety laws. This is the real junk science whose many examples illustrate the intensity and aggressiveness of the corporate campaign. To better know it when we face it, we have identified specific mechanisms that corporations use to deliver their commercial science: the media, politicians by campaign financing, the colleges and universities, obstruction of litigation (including SLAPP suits), corporate front groups, and such trade agreements as GATT and NAFTA. We need community-based scientific, engineering, and civic institutions to develop intelligent and honest research and to wisely apply scientific knowledge. Barry Commoner has helped set the standard for such sound science.

It is an honor to be here for Barry Commoner's eightieth birthday, and this celebration is a reflection of the integrity and advocacy of the global viewpoints that he has brought to his work over the years.

He is really one of the most complete scientists. You realize that when you read his books because when he talks about environmental issues and energy issues, he does not just deal with the particular risk levels or hazard levels or trends. He asks much more fundamental questions as to what is the utility of the petrochemical industry and why do we even have a fossil fuel-based industry projected into the next century? What is the nature of industrial organizations that has to be changed so that we develop different kinds of incentives for different kinds of environmentally benign technologies? None of these statements ever gets reduced to a sound bite, which, of course, disqualifies Barry Commoner from appearing on the evening's televised news.

The distortion of the media's motif of reflecting judgments and opinions by various personages in our society has now become a critical issue of sound-bite journalism, where only conclusions can be mentioned and, increasingly, conclusions that are not upsetting to the established powers. My remarks will focus on the phenomenon of junk corporate science and scientists, and how in almost an Orwellian sense the attack on scientists working in the consumer and environment and workplace safety and other areas has been orchestrated to put the shoe on the other foot. So, led by people like John Stossel of ABC News, they convert to corporatism, and Peter Huber, who wrote the book *Galileo's Revenge,* and Walter Olson, and others, the Manhattan Institute, the so-called American Tort Reform Association, and other corporate-funded groups have managed through the assistance of some systematically supported mass media to coin this phrase "junk science" and apply it in exactly the wrong arena. Instead of simply being on the defensive to try to ward off the criticisms of corporate junk science and scientists, they have been aggressive, on the offensive, in order to fulfill the principle that the best defense is an offense.

THE POWER GAME

The subject of corporate science, commercial science, must be viewed in the context of the power game. Just like bought-and-paid-for politicians and bought-and-paid-for media, bought-and-paid-for scientists are an instrument of corporate power. They are integral instruments in deferring the consequences of reality and warding off the mobilization of the citizenry and the application of regulatory health and safety laws. This is nothing new, but it is much more intense and much more orchestrated and much more on the offense.

I will give some illustrations of how long the corporations have gotten away with their type of commercially bought-and-paid-for science. Back when Southern California was suffocating under photochemical smog, the auto companies routinely denied that their emissions had anything to do with the air pollution situation in Southern California as a matter of adverse health effects. In 1950, Arlie Haagen-Smit, a professor of biochemistry at the California Institute of Technology, discovered that smog was produced by a photochemical process linked to oil refineries and automobiles. In response to Haagen-Smit, the petroleum industry tried to discredit his findings by funding research at the Stanford Research Institute that concluded that Haagen-Smith was wrong. By 1954, Stanford Research Institute's conclusion that smog was *not* linked to automobiles was the prevailing view, even after a special committee appointed by California's governor in 1953 affirmed Haagen-Smit's research and the Air Pollution Foundation, a group organized by civic leaders, established that auto exhaust was a major factor in Los Angeles smog. It would take years for the auto companies to admit that they were convinced. They finally did.

With its own well-financed scientists, the auto companies' strategy was to insist that the automobile's role be clearly proved and to construe any proof as narrowly as possible. And so, Professor Haagen-Smit was confronted by an industry's well-financed group of scientists and political operatives whose goal was to depict his finding as junk science. In fact, it was the auto and oil industries that were the purveyors of junk science and scientists, and years went by before Haagen-Smit's findings were translated into increasingly stringent air pollution control standards. Which means years went by while people were breathing these toxins and receiving the consequences: emphysema, cancer, and other ailments.

DPT VACCINE AS EXAMPLE

Another example of junk corporate science was the diphtheria-tetanus-pertussin (DPT) vaccine. As late as 1986, the American Academy of Pediatrics Committee on Infectious Diseases recognized the relationship between the pertussin vaccine and encephalopathy, and the necessity for some immunizations being contraindicated.

In order to avoid liabilities for injuries caused by the DPT vaccine, manufacturers sponsored research that dismissed the likelihood of neurological illness from the DPT vaccine. This research was paid for by Lederle Labs, one of the two U.S. manufacturers of DPT vaccine. In 1990, the *Journal of the American Medical Society* published the results of these paid researchers' findings concluding that there is no link between the DPT vaccine and seizures. When this research was published, the author failed to disclose his financial connection to the DPT vaccine manufacturer. Indeed, he also failed to disclose that he was a paid consultant to that manufacturer.

A third example which is very instructive is tobacco. As we all now know, the tobacco companies were engaged in a half-century effort to generate "scientific evidence" disputing any link between smoking and cancer, smoking and heart disease, or any other types of lung diseases. The research was cultivated and paid for by tobacco industry lawyers who represented the industry as an independent broker, as an independent institution, in order to protect it with the lawyer/client privilege. The July 1995 issue of the *Journal of the American Medical Association,* a classic issue, reported that tobacco company documents showed that by the 1960s Brown and Williamson had proved in their own laboratories that cigarette tar causes cancer in animals. When you see the thousands of medical journal articles and other scientific studies contradicting the tobacco companies' assertion over the years that smoking does not cause cancer or other ailments, you can see how formidable the propaganda machine has been by the tobacco companies. They even managed to deter the American Medical Association for a number of years by providing research funds after the Surgeon General's report in 1964 so that more studies could be made. It took quite a while for the AMA to begin challenging and going after the tobacco industry.

What is interesting about tobacco is that now it is very, very difficult for the tobacco companies to find one scientist to stand up and say—when they found *dozens* of scientists years ago—that tobacco is not connected to these diseases. That is always a good case study because it is a study of endless money buying endless "scientists" and research and then gradually shriveling down to fewer and fewer researchers parroting the industry line. If we trace the beginning of that decline, we are not able to say it began with the regulatory agencies, the Food and Drug Administration (FDA), other than the Surgeon General's report which came out of the same government department, and repeated Surgeon General reports from 1964 on. But it certainly did not come from Congress. It came from a much maligned propulsive lever, namely, private litigation. It came from the court law system, from a few people or next of kin finding contingency lawyers to file suit against the tobacco companies to try to prove that even though tobacco is consumed by the consumer, addiction has two players: the addictor and the addictee. The addictor's strategy was to go after youngsters and hook them into a lifetime of smoking when they are very impressionable and very vulnerable.

INTERNAL MEMOS, INCRIMINATING EVIDENCE

And so it was the production of internal company memos, from Philip Morris, R. J. Reynolds, and Brown and Williamson that led to the incriminating evidence that alarmed some members of Congress, like Henry Waxman, that nourished the evidential base for David Kessler at the FDA. And they got this entire latest stage of reaction and control against the tobacco industry underway. This is why the companies understand how powerful the court law system is and how important it is for them to destroy it under the guise of court reform and eventually make it very difficult for victims to have their day in court with full discovery powers and trials by civil jury.

Another illustration of junk science, or junk scientists, is the anti-depressant drug Merital, which is sold in the United States by Hoechst and was approved by the FDA in December 1984, and taken off the market in January 1986. The drug caused hundreds of allergic reactions, including fatal reactions such as hemolytic anemia, an immune reaction in which drug-induced destruction of red blood cells occurs. The company knew of a hundred cases of Merital-based induced anemia before the FDA approved the drug but failed to report any of them to the FDA in as prompt a manner as required, according to the Public Citizen Health Research Group.

Another example is Suprol, a pain killer manufactured by McNeil, removed from the market in May 1987. This drug caused acute kidney damage and a Congressional investigation chaired by the late Congressman Ted Weiss of New

York revealed that before Suprol was approved in the United States, its manufacturer had done studies showing that the drug caused kidney damage. When the Congressman raised the question as to whether or not McNeil Labs violated the adverse drug reporting requirements in not reporting these cases of kidney damage to the FDA more promptly, basically the answer was, "Well, we've taken it off the market." There is almost never a retrospective punishment operating in the area of the corporate people who are responsible.

Often in the assault on environmental and consumer and workplace safety science, there are arguments almost entirely by anecdote. You check the anecdotes of Peter Huber and others and they don't stand up. So there might be some people asking: "Am I arguing just by anecdote?" First, here are two books put out in the early 1980s by the Health Research Group of Public Citizen. One is called *Pills That Don't Work* and the other is called *Over-the-Counter Pills That Don't Work.* In *Pills That Don't Work,* there are 600 prescription drugs that have no evidence of effectiveness for the purposes for which they are advertised. They represented one out of every nine prescriptions in 1980, more than a billion dollars a year. They are all vouched for by drug industry scientists, drug industry representatives to doctors and hospitals, and *none* of them has demonstrated any evidence of effectiveness, which is required under the 1962 Drug Amendment Act following the thalidomide tragedy. Most of them are now off the market.

But just consider the amount of junk corporate science that's embedded in these pages, hundreds of prescription drugs, more than 600 prescription drugs. They are finally off the market. The book about over-the-counter pills that don't work deals with ineffective pills that are advertised for treatments of colds, pain, insomnia, coughs, allergies, losing weight, constipation, hemorrhoids, and other problems. So what we are really dealing with here is the institutionalization of commercialized junk corporate science that expands its tentacles in all directions: local, national, and global.

EFFORT TO DEBUNK IDEA OF GLOBAL WARMING

And speaking of global, the controversy now over global warming is moving toward an increasing consensus that, indeed, it is occurring and it is associated with the activities of human beings, especially the burning of fossil fuels. And yet, again, the coal industry, the oil industry, the auto companies and other industries, with the exception of the insurance industry which is beginning to wake up on this issue, have hired their usual scientists to attempt to debunk what is now an emerging consensus among scientists worldwide. Some of these challenges by corporate junk science have a modest redeeming value: they keep the assertions of the other side under some kind of scrutiny. The trouble is that most of the corporate junk science advocacy is not in the normal pattern of peer review. It is in the pattern of polemical *Wall Street Journal* type editorials. Therefore, its futility as a consistent cross-examination tool is de minimis.

There is a very well coordinated attempt to further defer response to the global warming phenomenon. And this attempt basically tries to rupture the causation between fossil fuel emissions and an increase in the temperature of the planet which, of course, has some pretty deadly consequences right down to the bottom of our oceans. The critique of the global warming position misses one of the major points, which Barry Commoner has often pointed out, and that is in responding to the global warming warnings.

Let's assume for purposes of this argument that we do not have 100 percent certainty yet. In responding, another argument must be made. That is that some of the mechanisms to reduce the impact on the global temperature are the mechanisms that will make energy efficiency a more frugal strategy. So we want to reduce energy use for economic efficiency reasons, for consumer frugality reasons. We want to reduce it because of recognized damage to the physical environment here on earth. We want to reduce it because we should not have to enter into geo-political wars and conflicts over fossil fuel reserves. And there are other more earthly reasons, such as balance of payments, deficits, why we should do it. One of the by-products, of course, is to relieve pressure on the global warming phenomenon. John O'Connor is co-author of a new book called *Who Owns the Sun* which illustrates this conflict between solar renewable energy that is decentralized and out of the control of the big oil and coal and nuclear companies and the desire for highly centralized, capital-intensive, corporate-dominated fossil fuel and nuclear industries to maintain their hegemony.

HOW CORPORATIONS DELIVER JUNK SCIENCE

One might ask, What are the mechanisms by which this kind of commercialized science gets delivered? Those of you who worked in the areas of lead and asbestos, autos and many other products, understand very well what the mechanisms are. They are not simply casuistry, not simply specious arguments, not simply diversions. They are connected as instruments of power to a number of institutions in our society that are contaminated by the grip of commercially distorted science. Just consider how long it took, for example, for the issue of lead toxicity to prevail. And still it is being challenged by the lead industry. Or how long it took for asbestos as a cause of abestosis and mesothelioma to be established, not simply scientifically but as an operating public health warning, leading to countervailing action.

There is an interesting and amusing sequence in the way companies respond to an alert that one of their products is hazardous, and in a recent publication by the Center for Public Integrity called *Toxic Deception* there is a nice chapter, "Science for Sale." I want to read you the sequence about how the asbestos industry argued its case until it was essentially defeated. This is brought together by Boston University's David Ozonoff, at the public health school.

"Assertion: Asbestos doesn't hurt your health. Ok, it does hurt your health but it doesn't cause cancer. Ok, asbestos can cause cancer, but not our kind of asbestos. Ok, our kind of asbestos can cause cancer, but not the kind this person got. Ok, our kind of asbestos can cause cancer, but not at the doses to which this person was exposed. Ok, asbestos does cause cancer, and at this dosage, but this person got his disease from something else like smoking. Ok, he was exposed to our asbestos, and it did cause his cancer, but we did not know about the danger when we exposed him. Ok, we knew about the danger when we exposed him, but the statute of limitations has run out. Ok, the statute of limitations hasn't run out, but if we're guilty we'll go out of business and everybody will be worse off. Ok, we'll agree to go out of business, but only if you let us keep part of our company intact and only if you limit our liability for the harms we have caused."

PAID SCIENTISTS BUTTRESS THE CASE

That's the story, isn't it? All along, paid scientists buttress this kind of case. The automobile area—that is one with which I am a little familiar—illustrates a more bizarre approach. The industry used engineers more than they used physicists. I remember once, in the 1960s, going to a technical conference on automotive safety engineering, one of the first ever, and a chief engineer for safety was there from General Motors. The issue of seatbelts came up. He got up and very seriously said that the technical bases at General Motors—implying that there is no greater level of confidence in this area in the world—have concluded that seatbelts were inadvisable because at above-30-mile-per-hour impacts, the internal organs of human beings could not withstand the level of deceleration anyway. So that even if the passengers were restrained, their internal organs would rip out in the deceleration by the G-forces.

An Air Force colonel, John Paul Stapp, decided that he had to disprove that. At one of the Air Force bases, based on a Pentagon program which came after a realization by the Pentagon that it was losing more Air Force pilots on the highways in the United States than in the Korean War, he strapped himself into a sled and went up to 180 miles per hour and then decelerated and proved that the human body would take a deceleration well above 60-mile-an-hour impacts on the highway. That's what he had to do—with some damage to his retina, I might add—to disprove General Motors' junk science. It was very noble. Stapp was a medical doctor in the classic tradition of the finest the medical profession has to offer.

For years, the auto companies tried to defy the laws of physics in so many ways and tried to block any attempts to translate contrary findings into safety standards.

The nuclear industry—some of you have had some semi-amusing conversations about it. For instance, how many of you have met nuclear physicists and struck up a conversation and said, "Well, you know, you really should pay attention to solar energy because nuclear energy is too expensive; it's too hazardous from a national security point of view; it's vulnerable to earthquakes more than most forms of energy; it's the only form of energy that cannot get private insurance; it has to be given limited liability by the federal government under the Price-Anderson Act; and you haven't figured out a way to store radioactive waste yet. Why don't you pay some attention to solar energy?"

One physicist told me that the reason why nuclear physicists were not interested in solar energy is because solar energy was just sophisticated plumbing and not enough of an intellectual challenge, compared to the atom.

Anyway, how many times have you heard nuclear physicists say, "The problem with solar energy is it's too diffuse and you just can't store it." If there is a physics exam in the eighth grade and a student gave that answer, what would the teacher say? But I just heard it three weeks ago when I sat on a plane next to a nuclear physicist.

I looked at him and I said, "You cannot be serious."

He said, "Oh yes, I am."

"You mean it's too diffuse?" to which I say, "Do you remember when you were a kid and you didn't like ants and you picked up a magnifying glass outdoors and fried the ant? Why did you forget that lesson?"

THE DELIVERY SYSTEM

Whether it's toxins or something else, Barry Castleman and Peter Montague and others here have felt the brunt of the delivery systems of corporate junk scientists and corporate junk science. Some day, we'll find a better title for that, but this is a tribute to Peter Huber in reverse, to use those phrases.

The first delivery system is through the media. There is a constant connection with *Forbes Magazine,* the *Wall Street Journal,* the trade association magazines, and the U.S. Chamber of Commerce. Any time an attack on environmental, consumer, or workplace science is unleashed, there are the automatic columns, the automatic editorials, and the very, very efficient disbursal of reprints to people at ABC, NBC, CNN, and various publications. And if the reprints do not persuade the members of the media, they do raise doubt. They do produce hesitancy and hesitancy in a fast-paced media program scene tends to result in, "Well, let's not have this person on the program. Let's not have this environmental scientist on the program."

The second delivery system is politicians by campaign financing. They have their high-profile elected representatives in Congress and elsewhere parroting their line, and you see it works very well with people like Dan Quayle, George Bush, Ronald Reagan, and many members of Congress. After all, it was Ronald Reagan who, while campaigning for president, said, "80 percent of all air pollution comes from trees," leading to some very nice cartoons. Ronald Reagan was so misled by corporate scientists that on two occasions in front of many members of the press, once while campaigning for president and once while he was Commander-in-Chief, made the following statement, "Nuclear missiles, once unleashed, can always be recalled."

It's not as if these agencies can establish overwhelmingly modest state-of-the-art standards, against industry, even if there *was* no Congress on their backs. The daily hammering by industry lobbyists, industry lawyers, and industry scientists, and the threat to close down the factories or go abroad, is enough to bring these proposed standards, if they were ever issued at a high level, down to extremely mediocre levels indeed.

THE COLLEGES AND UNIVERSITIES

The third delivery system is through the universities and through the colleges. An enormous amount of money is going into shaping research priorities, deterring other research projects. Major moonlighting by scientists and professors from Stanford to Harvard to MIT is a normal thing. As a matter of fact, among many Stanford professors, if they cannot double their salary as a professor by moonlighting, they are considered deficient in their entrepreneurial activities.

Corporate joint ventures with various departments in universities have led, in turn, to several consequences. One is that science on campus becomes secret rather than open, which is a severe violation of the traditions of scientific research. Science is funded by companies and becomes proprietary, and professors and graduate students working on these projects are bound not to discuss the findings. Indeed, the findings become, in more cases than not, the private property of the sponsoring biotechnology corporation or computer company.

Secondly, research priorities are twisted on campus because corporations are where the money comes from. Third, dissenters on campus are intimidated or downgraded or otherwise inhibited. You see now struggles for tenure by young professors who spoke out or who were expert witnesses in trials. They are very much on the defensive. The presidents and the Boards of Trustees of universities, who are already close to industry and commerce, are even closer when you have these secret joint venture agreements which are not disclosed, in many instances, to the students and the faculty, not to mention the various patent licensing arrangements that are spreading. A fourth impact on campus, still in need of inquiry, is to what extent tuition increases are cross-subsidizing the need for the universities to establish state-of-the-art laboratories and other investments in order to attract corporate research money in the bidding war. That is increasingly being looked into by some student groups from the University of Wisconsin and elsewhere.

OBSTRUCTION OF LITIGATION

Back to the delivery system: in addition to politicians, university-funded scientists, and the media, we have massive obstruction of litigation that seeks to establish scientific policy to protect wrongfully injured or sick people. In our new book, *No Contest, Corporate Lawyers and the Perversion of Justice in America* (Random House, 1996), we document in chapter after chapter how corporate law firms, on behalf of their clients, can destroy documents, withhold documents, obstruct, delay, harass witnesses, and engage in a variety of other judicially inappropriate or unethical maneuvers in part because the lawyers are paid by the hour and the more they delay, the more they obstruct, the more they complicate, the more money they make. The money they are paid with is from some of the richest corporations in the world, and it is all entirely deductible under the IRS code. So there is not exactly an incentive to economize and abbreviate these prolonged litigations.

Another delivery mechanism is SLAPP suits—Strategic Lawsuits Against Public Participation. There is a book out on them published by a professor at the University of Denver Law School that is available from Temple University Press in Philadelphia. There are hundreds of SLAPP suits filed against citizens and their specialists who are challenging developers, toxic waste treatment companies, incinerators, and other installations in the local areas where people live. For example, a group in Winona, Texas, took on U.S. Ecology, which once was Nuclear Engineering Corporation. It is one of the biggest toxic waste treatment companies in the country and it was going to build another installation in Winona, Texas, notwithstanding concern by the 500 people who live in that small town about rising ailments and toxic diseases.

The group started fighting U.S. Ecology, and in October 1996 the company, through its law firm, Latham and Watkins, a big one out of Chicago and other cities, filed a $35-million lawsuit against these people—just ordinary villagers and their supporters—including a racketeering provision, even though last March the company withdrew the project and said it wasn't going to build there. So I called up the law firm and I said, "Why are you persisting in this suit?" And indirectly I was informed, Well, the law firm wasn't excited about this SLAPP suit, charging defamation and interference with economically advantageous relationships and racketeering on the part of this humble citizen group. But, they said, the company was adamant about it.

SLAPPING BACK

Such suits are happening around the country. It is a very chilling prospect, not just to citizens, but to their scientific advisers. Indeed, I recently saw a letter from a large law firm to a potential expert witness who was cited by the defense to testify. It said basically that, If you testify, we're going to sue you. They even sent a letter to the Lawrence Livermore Lab on one occasion, saying that, We understand you are going to conduct tests, and we caution you about doing it. This is how bold they are getting. Of course, there is now, in the good old American tradition, an anti-SLAPP movement to generate slap-back lawsuits for malicious use of legal process, and a few big judgments in that area may sober up some of these billable hour law firms and their corporate clients.

Two more delivery systems: one is corporate front groups. There are now dozens of front groups with all kinds of fancy names that sound like they are environmental groups and consumer protection groups. They are spending a lot of money generating a counterattack against environmental, consumer, and workplace scientists. Not the least of them is the Manhattan Institute which received a glowing feature article in *The New York Times* (Spring 1997) in the Metropolitan section, without any indication that it is a front group, almost bought and paid for by major corporations which it solicits every year. The New York Public Interest Research Group—which has been around for more than twenty-five years, is all over the state, has produced all kinds of good research on toxins and incinerators, and is supported by tens of thousands of students with offices in twelve cities around the state—has never received a single feature article in twenty-five years by *The New York Times,* even though its findings are quoted regularly and its off-shoot groups are sources of information to *New York Times* reporters, like the Strap Hangers Group that improves New York subways. This is an illustration; the New York Public Interest Research Group does not have a high-powered public relations firm and the Manhattan-type groups do much better at getting into *The New York Times.*

These front groups are often very confusing to reporters who don't catch up to them for a while but take their releases as if their findings are impartial and objective and competently produced.

GATT AND NAFTA

The last delivery system may, indeed, be the most menacing. It's called GATT (the General Agreement on Tariffs and Trade) and NAFTA (the North American Free Trade Agreement). The mandate behind GATT is that the imperative of international trade is given supremacy over health and safety standards legally sanctioned or proposed in the respective signatory countries. This means that the countries that have higher environmental, consumer, and workplace standards are countries that are going to get in trouble in GATT because other countries that have lower standards and export products to our country, like fruits and vegetables that don't have to meet the same pesticide control standards that we have, can take the United States or California or any local jurisdiction to tribunals in Geneva, under the World Trade Organization, and these tribunals are kangaroo courts. They are secret. They have no public transcript, no standards of ethics for the trade judges who have no experience in environmental, consumer, and workplace usage—just trade judges—and no independent appeal. If the United

States loses, we have to repeal our law or pay perpetual trade fines to the winning country.

We've already lost to Venezuela on reformulated gasoline. In order not to pay the fines, we are reformulating the standard for reformulated gasoline. And we're about to lose to Mexico on the tuna/dolphin. The respective countries—the European common market, Canada and Japan—have listed hundreds of U.S. environmental and consumer protection laws that they think are GATT illegal. That is, they are trade restrictive. Countries that treat their people more humanely will get in trouble with GATT; countries that treat their people less humanely do not get in trouble with GATT. Under GATT, international trade in products made by brutalized child labor in countries like India, Pakistan, and Bangladesh are legal. They are legal. And any country that signs on to GATT, that then prohibits the importation of products made from brutalized child labor abroad, will be considered in violation of GATT and will have to pay perpetual trade fines pursuant to a secret tribunal decision that is a foregone conclusion (given the technical statements in the GATT agreement that no country can ban products based on how they are produced except when they are produced by prison slave labor).

CODE WORDS

The words "sound science" go throughout the GATT deliberations, and "sound science" is a code for mainstream corporate science, the kind of science that dominates the Codex Alimentarius, which is the food safety standard arm of the United Nations, and dominates the advisory committees to GATT. And "sound science," whenever you hear the phrase, is basically a semantic camouflage for the corruption of science by money at the behest of corporate policies that are increasingly lawless and increasingly bold and aim to dismantle our democracy step by step, whether it is access to the judiciary, whether it is getting honest science prevailing in public policy, or whether it is getting our legislatures and regulatory agencies to do the jobs that they were supposed to do to apply the relative certitudes of contemporary science to the health and safety of our workplace, our marketplace, and our environment.

It is fitting, in conclusion, that the standard for sound science and basic analysis is the man whom you are honoring today. And, it is also fitting to indicate how much work we have left to do, in order to mobilize and counterattack with broad talent all over the country the trends and the phenomena that I have just briefly described. The kind of counterattack has to be rooted in understanding what was illustrated in a statement by Jean Monnet, one of the founders of the European community, who said, "Nothing is possible without people, and nothing is lasting without institutions." And while one may certainly view Barry Commoner as an individual, and certainly may use a metaphor by describing him as an institution,

I'm sure that he would be the first to agree that we need community-based scientific, engineering, and civic institutions all over our land so that we can develop the counter-building forces of intelligent and honest research and apply them to the problems at hand. Otherwise, corporate science can become one of the major menaces of our world.

CHAPTER 5
Barry Commoner's Science: An Anecdotal Overview

Danny Kohl

Barry Commoner's scientific career is best characterized by his insistent commitment to holistic (as opposed to reductionist) approaches to understanding how living things function and his alertness in bringing the most modern tools from physics and chemistry to bear on the properties of living systems. The pioneering work of his laboratory on the life history of tobacco mosaic virus was widely admired. In addition, his was the first work which used a magnetic resonance technique to investigate biological phenomena. Characteristically, these studies utilized whole, living, functioning organisms. He pushed the limits of sensitivity for measuring small differences in the nitrogen isotopic composition of drainage water to investigate the relative contributions of fertilizer N to the high nitrate levels found in an agricultural watershed. While the conclusion that N applied as fertilizer was responsible for about half of the nitrate pollution initially met with fierce resistance, the methods which he conceived are now widely used. His ideas about the multiple roles of DNA in inheritance were still less warmly received. While being involved in all of this, the Barry whom many of you know best found the time and energy to be a major figure in bringing the dangers of radioactive fallout to public attention, to be a vigorous and effective opponent of the war in Vietnam and to play a leading role in establishing the environmental movement.

While you all know something of Barry Commoner's contribution to the environmental movement, I would like you to know about his contributions to laboratory science. Before I do that though, I'd note that Barry has been extremely successful in all variety of tasks he has undertaken. One of Barry's assigned tasks during World War II was to devise an apparatus which would allow an airplane to spray the new miracle compound—DDT—in the jungle in an attempt to protect our troops from malaria. He succeeded.

After the war, but still in the Navy, Barry was assigned to the office of the senator from West Virginia, Harley Kilgore. Senator Kilgore was an important player in one of the major debates of that time: namely, who would control atomic energy, the mammoth new force which had been unleashed on the world? Would it fall under military control? If it did, the assumption was that the emphasis would be on weapons. Or would it fall under civilian control? In this case, some believed—although I don't know how Barry was articulating the future at that time—that the emphasis would be on peaceful uses, principally producing energy so inexpensively that it would be too cheap to bother metering.

Obviously, Barry was on the side of civilian control of atomic energy, as all of us in this hall would have been. And our side won. Our prize? The Atomic Energy Commission—soon to be better known as the "Hated AEC." Ah, but consider the alternative! I suspect you will hear much more about the AEC during this symposium.

After his wartime service, Barry came to Washington University in St. Louis where his main focus was on studies of tobacco mosaic virus (TMV). This was an extremely shrewd choice for someone who had been interested in the Secret of Life since grade school. It is a very regular rod, 300 nm long and 15 nm in diameter, composed of identical protein subunits and an RNA (no DNA) molecule. TMV was arguably among the simplest of life forms, although Barry was always quick to point out that it could not replicate outside of an intact living cell.

The annual meeting of the American Association for the Advancement of Science (AAAS) is one of the largest gatherings of scientists. The AAAS gave a prize each year—called the Cleveland Prize—for the best paper presented at its annual meeting. In 1953, Barry won the Cleveland Prize for a paper on TMV. In an act very typical of Barry, on his return to St. Louis, he split the prize money among the folks in the lab who had done the day-to-day work in producing the data that went into that paper.

By the middle to late 1950s, all biology having to do with the replication of TMV—as well as many other areas of the biochemistry of inheritance—took place in the context of the Watson/Crick model of inheritance.

Watson and Crick published a paper in *Nature* in 1953 (Vol. 171, pp. 737-738) proposing that deoxyribonucleic acid (DNA) was a double helix (Figure 1) containing a sugar/phosphate backbone with one of four compounds called bases attached to each sugar. The next paper in the same issue of *Nature* made another important contribution to our understanding of the structure of DNA. Its authors were M. H. F. Wilkins and his colleagues. A. R. Stokes and H. R. Wilson, which was followed by a third hugely important paper by Rosalind Franklin and her colleague, R. G. Gosling. Wilkins later shared the Nobel Prize with Watson and Crick. Franklin died at thirty-seven, before the Prize was awarded.

Watson and Crick published a second paper in a later issue of *Nature* that same year which discussed the genetic implications of the double helix.

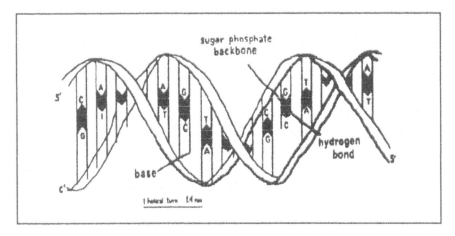

Figure 1. DNA double helix.

Watson and Crick had formulated what they called "The Central Dogma of Molecular Biology"—imagine in science calling something a *dogma*! This designation infuriated me as a starry-eyed graduate student with many idealistic notions about Science. (It still does.)

The message which the Central Dogma was meant to convey is spelled out in the text accompanying the figure (Figure 2) in the 3rd Edition (1976) of Watson's influential book, *The Molecular Biology of the Gene*. Two footnotes stated that one-way flow of information, the defining feature of "The Central Dogma," is not always the case. Despite this, Watson and the molecular biology community continued to use the designation, "The Central Dogma of Molecular Biology."

As noted, embedded in the Watson/Crick formulation was the idea that DNA was a self-replicating molecule. Barry took considerable exception to this, insisting that no biological unit less complex than a cell was capable of self-replication. In addition, Barry proposed that certain proteins, as well as DNA, played a role in biological inheritance. Finally, he proposed that DNA played more than one role in inheritance. In addition to the DNA in the euchromatic regions being the stuff of genes, Barry hypothesized that the amount of DNA "sequestered" into the heterochromatic region served as a coarse governor of the basal metabolic rate (Figure 3).

The central feature of this "sequestration" theory used the facts that DNA was an extremely stable molecule and the nucleotides which provided the bases in DNA were also crucial co-factors in metabolism. So, the argument went, if these nucleotides were sequestered into stable DNA, they would not be available to power metabolism. I was not a fan of this theory because, as I argued with Barry, the diagram needed one more arrow, an arrow indicating that ATP (adenosine

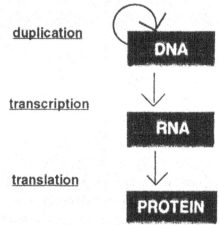

> "Here the arrows indicate the direction of transfer of the genetic information. The arrow encircling DNA signifies that it is the template for its self-replication; the arrow between DNA and RNA indicates that all cellular RNA molecules are made on DNA templates. Correspondingly, all protein sequences are determined by RNA templates. Most importantly, the last two arrows are unidirectional, that is, RNA sequences are never copied on protein templates; likewise, RNA never acts as a template for DNA." (a)
>
> a. Watson, J.D. *Molecular Biology of the Gene* (3rd Edition), W. Benjamin, Menlo Park, CA, Page 209, 1976.

Figure 2. The central dogma of molecular biology.

triphosphate) and GTP (guanosine triphosphate) can be synthesized, as well as sequestered.

Nonetheless, it led to a testable prediction. Barry predicted that organisms, like lungfish, that had lots of DNA, much of which contained few if any genes, would have slower basal metabolic rates than did organisms which had lesser amounts of DNA per cell.

Testing the predictions of this theory became Jim Bennett's thesis topic. To do so, Jim had to draw blood from lots of animals in the zoo. This required both skill and daring. Have you ever tried getting blood from a hummingbird or from a very bad-tempered ostrich? Fortunately for Jim, he located an animal dealer in Florida who provided blood from water snakes and alligators. And fortunately for me, it was Jim's project rather than mine.

Three papers outline Barry's views on the role(s) of DNA [1-3]. To put it mildly, these papers were not well received by the emerging community of molecular biologists. Unfortunately—or perhaps fortunately—there is no time to

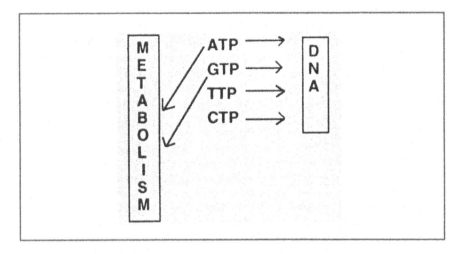

Figure 3. Commoner's sequestration theory.

go into this controversy in detail. But it does lead to an illustration of a lesser known feature of Barry's personality. Barry was extremely tolerant of disagreement, at least out of the mouth of this smart-assed graduate student.

Given the controversial nature of Barry's point of view, a debate was arranged at an amphitheater at Washington University Medical School. I thought Barry's presentation was unnecessarily confrontational that night, so when I got called on in the question period which followed, I said: "Professor Commoner, isn't it true that your position can be restated as a testable hypothesis, as follows?" And I proceeded to lay out the Sequestration Theory. Barry laughed and said: "In case anyone doesn't know who is asking that question, it is Dan Kohl and he is a graduate student in my lab. What he is really saying is: 'Barry, I don't like the way you put the question. It would be better if you said it my way.' " To which I replied: "That's right. And if you'd spend more time in the lab, I wouldn't have to come to places like this to talk with you." Barry thought that was very funny. I would have fired me for provocative rudeness!

Permit me now to sketch for you briefly another area in which Barry made an enormous contribution. It also illustrates that he was devoted to bringing the most sophisticated available approaches to studies of biological processes.

A word of background. The metabolism of all cells includes two categories of processes: the breaking down of pre-formed molecules to provide energy for cellular and tissue functions, such as the motility of sperm or the contraction of muscle, and the synthesis of more complicated molecules from simpler parts. The breakdown of, for example, sugar molecules, which is called *oxidation,* involves the loss of two electrons at each step. But the electrons are transferred one at a time. In organic molecules, electrons occur in pairs. When the first electron is lost, the organic molecule has one, unpaired electron. Chemists call organic molecules with an unpaired electron free radicals. In 1954, it was already known that free radicals played an important role in metabolism, among other important biological processes. But no one had a way of measuring them directly.

In his August 1958 article in *Scientific American,* George Pake tells the story leading up to the first direct measurement of free radicals in living biological material. The topic was considered important enough to command the cover picture of *Scientific American.*

Pake wrote:

> After attending a chamber music concert, a few members of the Washington University faculty repaired to someone's home for coffee, and the conversation turned... to shop talk. Barry Commoner of the botany department fell to discussing the theory that free radicals play an important role in the processes of oxidation and reduction in living cells, and he remarked how difficult it was to detect free radicals in living systems. I suggested that electron resonance might be helpful and offered to help Commoner and his group learn how to use the method. Thus began a most interesting series of experiments [4, p. 66].

Pake went on to note that it took the innovative skills of Jack Townsend of the physics department to design a machine that would tolerate the water which is ubiquitous in living biological systems.

What then is the basis for the technique Pake brought to Barry's attention?

Accompanying Pake's article was a drawing in three panels of a toy top spinning. The top is, of course, interacting with the earth's gravitational field. The result is that the top precesses around the direction of the gravitational field. The frequency of the precession is governed by *both* the magnitude of the gravitational force and the angular velocity at which the top spins. In the middle panel, an electron is depicted in the center of the cartoon. Electrons have a quantum mechanical property called "spin" which is analogous to the spinning top. Since the electron is an electrically charged particle, its motion (spin) leads to the generation of a magnetic field. If the spinning electron is subjected to an external magnetic field, as indicated by the arrow, it precesses just as does the top. As with the top, the rate of this precession is determined by the angular velocity of the spinning electron and the strength of the externally applied field. A small piece of magnetic iron placed in a magnetic field can be displaced at any angle from the direction of the applied magnetic field. If you want to rotate it more, simply apply more force. One of the important insights of quantum mechanics is that the magnetic moment associated with electron spin can be aligned only parallel or anti-parallel to the direction of the applied field. Physicists knew from theory that if a second external field, perpendicular to the main magnetic field, were applied, the spinning electron would absorb energy from this field when the frequency of the applied field is equal to the frequency of the electron's precession. This would result in the electron's spin reversing direction, the associated magnetic movement now pointing antiparallel to the applied field.

Pake includes in his article a diagrammatic representation of the instrument Jack Townsend built to detect free radicals in biological systems. An oscillator sends energy at 10,000 cycles/sec (3 cm wavelength) down a wave guide where it impinges on a sample. The strength of the magnetic field is varied, resulting in a change in the rate of precession of the electron spins' magnetic moment. When the energy necessary to flip the electron spins is just equal to the energy supplied by the oscillator, energy is absorbed as the spins flip over. This is observed as a dip in the energy which arrives at the receiver. In an attempt to convey the underlying physical basis of the technique, it is called electron spin resonance—ESR for short.

A 1954 paper in the journal *Nature* by Barry, along with Townsend and Pake, is *the very first paper which reported direct observation of free radicals in biological systems* [5]. In this paper, it is clear that Barry understood the broad potential applications of ESR for the study of biological systems. In this paper, Barry wrote: "Free radicals have been implicated as decisive participants in (a) biological oxidation-reduction reactions, (b) the action of ionizing, ultraviolet and visible radiation in biological systems, (c) chemical carcinogenesis. Although these inferences, if correct, are of far-reaching importance, all have been derived from indirect or model experiments; free radicals have not (*until this time*) been demonstrated in living things." In the next decade, Commoner and his colleagues produced a remarkable series of papers covering all of the topics which he enumerated in that first paper [6-12].

The second paper [6] (which is the one that attracted my attention when I was an undergraduate student) reports studies of photosynthesis; the third and fourth [7, 8] report studies of oxidation reduction systems. Then two papers co-authored with a surgeon [9, 10] led to a tool for diagnosing liver disease. One [11] is my first paper. The "fast kinetics" which it reports were on the millisecond time scale. We were very impressed with the experiments this early computer allowed us to do. Now, events which take place in an interval ten orders of magnitude faster are the subject of study.

Seeing the title again reminded me that I was extremely reluctant to begin laboratory work. Barry would come into the lab and say, "Dan, when are you going to start doing experiments?" I told him not to worry, that I would start soon but, at the moment, I did not know enough. Finally, I bit the bullet and went into the lab. My colleague, John Heise, had been showing me how to tune the machine and record a spectrum. John left the room momentarily, leaving me staring intently at the instrument trying to remember everything John had just told me. At that moment, Barry walked into the lab accompanied by Hans Gaffron, one of the giants of research directed toward understanding photosynthesis. (I was in awe of Gaffron.)

Barry said: "Dan, would you please show Professor Gaffron how to record a light induced free radical spectrum from the green algae you have in the sample chamber?"

My heart in my mouth, I fiddled with the instrument. When it became clear that no spectrum would be forthcoming soon, Barry asked rather gently: "Dan, could you use some help?" And he stepped forward, tuned the instrument and ran the sample.

Gaffron, with a twinkle in his eye, said, in what was meant to be a compliment, "Barry, I wouldn't have thought you knew how."

While it is true that most senior professors know *about* methods and techniques, they usually have people in the lab who know how to use the methods and techniques which the professor only knows about. Not Barry. At least not in this case. Barry not only knew *about* ESR, he also knew how to get the instrument to produce data.

Notice something else about my first paper. The authors are Commoner, Kohl, and Townsend.

When the paper had been written, Barry said: "I think I'll put the authors in alphabetical order. What do you think, Dan?"

I replied: "What if my last name were Abrams?"

Barry: "I'd think of another reason." Actually Barry had every right to be the first author of that paper. By the time of the last listed paper [12], I was playing enough of a role in the research to deserve to be first author.

Finally, it did not strike me until I was preparing this presentation that almost all of those papers appeared in the three most prestigious journals: *Science, Nature, Proceedings of the National Academy of Sciences.* Somehow, I was not alert enough to internalize this teaching by example. My subsequent papers appeared in more pedestrian journals.

Another of the stories connected with this work involves the occasion on which Barry was invited to Minnesota in the late 1950s to give a seminar entitled, "The Role of Free Radicals in Biology." The story goes—Barry will have to tell us whether it is true or apocryphal—that the talk was picketed. Among the signs the protesters carried were: "Radicals have no role in biology or anywhere else" and "Radicals should not be free. They should be in jail."

Still another major scientific area in which Barry was engaged flowed directly from his interest in environmental problems. Barry had become concerned with the pollution of surface waters with nitrate. He had noted the correlation between the increase in nitrate in surface water and the use of N-fertilizers in agriculture. As he tells it in his well-known book, *The Closing Circle,* one day the phone rang and it was Leo Michl, the Director of Public Health in Decatur, Illinois. Decatur is in the heart of some of America's richest farm land, in the middle of the Great American Corn Belt. Decatur takes its drinking water from an impoundment of the Sangamon River. The Sangamon drains a 900-square-mile agricultural watershed dominated by the row cropping of corn and soybeans. Michl told Barry he had read of Barry's concern about water pollution, and that, he—Michl—had a problem. Every spring, the nitrate concentration in the Sangamon River exceeded the U.S. Public Health Service's limit of acceptability and he wanted to know where that nitrate was coming from. Specifically, he wanted to know what fraction of that nitrate was the result of the addition of fertilizer N to corn in the watershed. He further wanted to know whether Commoner was just a talker or would he take on a practical problem which Michl perceived to involve the public's health.

It was as if Leo knew Barry well and knew exactly which button to push. The challenge was irresistible. The only question was how to proceed. At this point, Barry writes in the book, he remembered the decades-old observations in his laboratory that different components of the N cycle had slightly, but measurably, different concentrations of ^{15}N, the heavy, *stable* (not radioactive) isotope of N.

The lighter, stable isotope, ^{14}N, accounts for 99.6337 percent of the N atoms found in the atmosphere in the form of N_2, nitrogen gas. The heavier isotope, ^{15}N, is 0.3663 percent abundant in the atmosphere.

As a result of their difference in mass, a compound containing ^{14}N tends to undergo chemical reactions at a slightly higher rate than does the same molecule bearing an ^{15}N. This leads, for example, to the N-containing compounds in manure or composted garbage having higher ^{15}N abundances than the N in air, as indicated in Figure 4. By contrast, the N in ammonia (NH_3) fertilizers made by the Haber-Bosch process at high temperature have an ^{15}N abundance which is very close to that found in air. This is a consequence of atmospheric N_2 being the source of the N and the high temperature at which it is carried out. While the differences in the ^{15}N contents of garbage and commercial N fertilizers are small (see Figure 4), with care, and the best then available instrument, they were measurable.

Barry bet that, using this difference in ^{15}N abundance, he could detect the time at which grape growers in France switched from Paris garbage to commercial fertilizers for fertilization of their vines. To test the idea, Barry acquired French champagnes of various vintages. He was right. Since it took only a relatively small volume to make the measurements, and since to waste is a crime, there were a good many happy campers.

It was these observations which Barry pulled from his memory and which he decided to use to answer Leo Michl's question. The two potential sources of nitrate-N in that watershed were N fertilizer and soil-N made soluble in water which percolated through the soil by processes inevitably associated with growing crops. (In that watershed, there was no significant sewage outfall or livestock industry.) As a zero-order approximation, all we had to do was measure the ^{15}N content in Decatur's water and compare it to what we would anticipate if it all came from soil N on the one hand vs. fertilizer N on the other. The closeness of the ^{15}N in the water to that in the two sources (with the amount of ^{15}N in fertilizer

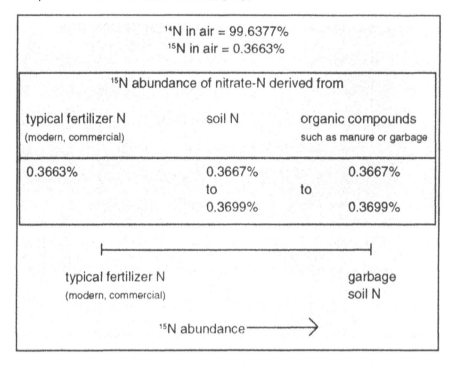

Figure 4.

N being less than that in soil N) would indicate the relative contribution of the two sources to the nitrate-N in the water.

Georgia Shearer was involved from the beginning. I first heard about the study in detail shortly after its inception. It was on an anti-Vietnam war occasion. We at Washington University, with Lindsay Mattison being the point man, had played a major role in the publication of *The Politics of Escalation,* a book which correlated every major escalation of the war with the danger to the hawks posed by the opportunity for a negotiated peace. Barry and I had been to see Harold Gibbons, an anti-war leader of the Teamster's Union, to ask him for $5,000 to help with pre-publication publicity. He gave it to us. On the way back to the university, Barry told me about the nitrate study and invited me to play a role in it. Since I had just finished a piece of work and was trying to decide what I would do next, I was delighted to accept his invitation.

After making many thousands of measurements on samples of water within the Sangamon River watershed, we concluded that about 50 percent of the nitrate-N in the impoundment of the Sangamon River came from fertilizer N and about half from turning over the soil to grow crops on it, a finding we published in *Science* in 1971 [13]. Actually, if one were to criticize this study as one in which a cannon was used to kill a fly, one would have some claim to justification. At a cruder level, the amount of fertilizer N sold in that watershed was approximately the same as the amount one would expect to be produced by the conversion of soil organic N to soil inorganic N, a process called mineralization. Nonetheless, the response to this finding was ferocious.

While Barry stayed associated with subsequent related work for a time, his mind was more and more occupied by other things. Georgia Shearer and I inherited the work and have made most of our careers applying and refining this approach, which Barry pioneered, to a variety of problems.

While most know about at least some of Barry's contributions in the policy arena, I wanted to highlight Barry's seminal contributions to laboratory science. In all cases, he used his phenomenal peripheral vision to bring to bear just the right techniques from chemistry and physics—often the newest and most sophisticated techniques—to get answers to the important biological questions he wanted to investigate.

Happy birthday, Barry. And many more.

REFERENCES

1. B. Commoner, Deoxyribonucleic Acid and the Molecular Basis of Self-Duplication, *Nature, 202*, pp. 960-968, 1964.
2. B. Commoner, Roles of Deoxyribonucleic Acid in Inheritance, *Nature, 203*, pp. 486-501, 1964.
3. B. Commoner, Failure of the Watson-Crick Theory as a Chemical Explanation of Inheritance, *Nature, 220*, pp. 334-340, 1968.
4. G. E. Pake, Magnetic Resonance, *Scientific American, 199*, p. 58, 1958.
5. B. Commoner, J. Townsend, and G. E. Pake, Free Radicals in Biological Materials, *Nature, 174*, pp. 689-691, 1954.
6. B. Commoner, J. Heise, and J. Townsend, Light-Induced Paramagnetism in Chloroplasts, *Proceedings of the National Academy of Sciences, 42*, pp. 710-718, 1956.
7. B. Commoner, B. B. Lippencott, and J. V. Passoneau, Electron-Spin Resonance Studies of Free Radical Intermediates in Oxidation-Reduction Enzyme Systems, *Proceedings of the National Academy of Sciences, 44*, pp. 1099-1110, 1958.
8. B. Commoner and T. C. Hollocher, An Electron Spin Resonance Analysis of the Mechanism of Succinic Dehydrogenase Activity, *Proceedings of the National Academy of Sciences, 47*, pp. 1355-1374, 1961.
9. B. Commoner and J. L. Ternberg, Free Radicals in Surviving Tissues, *Proceedings of the National Academy of Sciences, 47*, pp. 1374-1384, 1961.
10. J. L. Ternberg and B. Commoner, Clinical Application of Electron Spin Resonance, Differential Diagnosis of Jaundice, *Journal of the American Medical Association, 183*, pp. 339-343, 1963.
11. B. Commoner, D. H. Kohl, and J. Townsend, Fast Kinetics of Unpaired Electrons in Photosynthetic Systems, *Proceedings of the National Academy of Sciences, 50*, pp. 638-644, 1963.

12. D. H. Kohl, B. Commoner, J. Townsend, H. L. Crespi, R. C. Dougherty, and J. J. Katz, Effects of Isotopic Substitution on Electron Spin Resonance Signals in Photosynthetic Organisms, *Nature, 206,* pp. 1106-1110, 1965.
13. D. H. Kohl, G. B. Shearer, and B. Commoner, Fertilizer Nitrogen: Contribution to Nitrate in Surface Water in a Corn Belt Watershed, *Science, 174,* pp. 1331-1333, 1971.

CHAPTER 6
Barry Commoner and the Hamburger Story: Can Ideology Prevail Over Science?

Piero Dolara

In the Center for the Biology of Natural Systems at Washington University in 1976, young researchers were studying the fact that the bacterial mutation rate was higher in the Ames test after the addition of liver microsomes. The effect was due to contaminants present in beef extract and in fried hamburgers as well. The communication of these results was followed by a tremendous popular and press response. The fact that potent mutagens were present in cooked meat raised the possibility that they could play a role in the induction of cancer. This prompted greater scientific interest on the topic increasing scientific articles from a few to several hundred per year. The original observation was then forgotten and after twenty years heterocyclic amines, formed in hamburgers during frying, are known to be involved in the induction of the most common human cancers.

I met Barry Commoner in 1976 in Florence. He was giving a series of speeches in Italy on the environmental crisis, explaining how ruinous it would be to rely heavily on nuclear power and on non-renewable energy resources. I knew from common friends about his work on free radicals and environmental pollution. During that lecture in Florence, I was immediately struck by his brilliant analyses and eloquence which reminded me of the approach of Socrates and Galileo to practical and philosophical problems. As a young scientist interested in toxicology, I asked him if he would have me as a guest scientist in America. "We never had a medical doctor in our team, why not!" was Barry's reply. The next year I got a fellowship from the National Institute of Health and moved to Saint Louis.

The Center for the Biology of Natural Systems (CBNS) was certainly an interesting place back in 1977, with a relatively small group of talented people working with Barry on exciting topics: energy, nitrate water pollution, appropriate technology for developing nations, scientific information, organic farming. Next to them was a more conventional group of laboratory scientists who were applying bacterial mutagenesis techniques to the analysis of known and unknown environmental pollutants.

DETECTING MUTAGENS

In those years, a biologist from the University of California at Berkeley, Bruce Ames, had devised a method for detecting mutagens by means of a simple bacterial mutagenesis assay—the salmonella microsome test—in which traces of mutagenic contaminants could be detected by measuring the mutations induced in selected strains of salmonella. Bruce Ames had the bright idea of incorporating in the assay a preparation obtained from rat liver, which could perform complicated chemical reactions on a bacterial plate. In this system, mutagens could be activated and inactivated in a way that mimicked the complex series of reactions taking place in a live mammalian liver. Hundreds of compounds had been tested with this methodology, showing that there was a close correspondence between molecules with mutagenic activity in the mammalian-microsome assay and substances that had carcinogenic activity in experimental animals and humans. Hence Barry Commoner had a simple but ingenious idea: use the salmonella microsome assay to detect mutagenic/carcinogenic chemicals in the environment. Barry had convinced the Environmental Protection Agency (EPA) of the potential usefulness of applying bacterial mutagenesis as a way of checking environmental contamination in water, air, and in connection with industrial production.

When I arrived in St. Louis in 1977, the team of researchers working in the lab was studying a peculiarity of the salmonella microsome assay—the fact that the bacterial mutation rate was slightly higher after the addition of the liver microsomes without any added chemical—a phenomenon which Barry had named "the microsome effect." Since the microsomes had been obtained from rats treated with Polychlorinated Biphenyls (PCBs), one hypothesis suggested that PCB might be involved in the increased mutation rate after the addition of microsomes. But PCB at similar concentrations was not mutagenic in the assay. Barry and his co-workers had discovered that some components of one of the ingredients used for growing bacteria—beef extract—contained contaminants capable of increasing the bacterial mutation rate.

By talking on the phone with people from the supplier of the beef extract, Barry had learned that the product was produced with a very simple procedure: boiling down beef stock obtained from a slaughterhouse. Beef extract was a very dark and gooey paste, which was diluted in water in the lab to make media for growing bacteria. Several people in St. Louis were trying to understand what component of beef extract was actually mutagenic.

HAMBURGER CONNECTION

As an Italian born just after World War II, I was not used to fast food and to hamburgers. Going to the local cafeteria, I was always amazed at how many hamburgers Americans were eating. I remember that in the university cafeteria an impressive looking black woman was grilling hamburgers on a large metal hot plate, scraping with the spatula a thick layer of dark material oozing from the beef patties on the grill. A local person might not have noticed it, but I was struck by the resemblance of this material to the beef extract and kept thinking about it.

One weekend that winter, Barry was out of town for a speech and left early on Thursday. I took one of those hamburgers to the lab and, after having extracted it with methylene chloride—the same procedures used for analyzing beef extract—I gave the residue to a lab technician for testing with the salmonella-microsome assay.

When I came back to the laboratory the following Monday, the technicians, who were all excited, showed me a series of plates with a very high number of mutant colonies and asked what was the chemical that I had tested. They could not believe that it was not a chemical, but a hamburger patty!

Barry came back on Tuesday morning, and when I showed him the plates, he said: "Oh, my God, what will people say after this . . ." We knew we had something important on our hands and all of us worked hard on this new development. By March, we were ready to present the data at the national meeting of the American Microbiology Society in Las Vegas and I was to be the speaker.

At Las Vegas, the meeting hall was full. The presence of potent mutagens in hamburgers was a title which spoke for itself. But what did it was the presence of major newspapers and television networks. My English was rudimentary, but the message was clear enough: some possibly carcinogenic compounds were formed during the cooking of meat at ordinary temperature, raising the possibility that cooked meat might play a role in the induction of cancer.

In the same period, we had sent the results to *Science* magazine. One of the referees clearly did not like Barry Commoner. He/she wrote an excessively nasty review of the paper, distorting the scientific evidence and finding non-existing technical and scientific errors. The editor could not but agree with our rebuttal. Given the importance of the topic, the article was published with no delay [1].

IN THE EYE OF A STORM

We found ourselves in the middle of a storm. Nobody at CBNS was very excited about the new developments. After all, we intended to investigate nasty chemicals and chemical pollution. What do hamburgers have to do with all this? On the other hand, the scientific evidence was there: how could we ignore the fact that potent mutagens/carcinogens are present in our daily food?

The "hamburger battle" was not a "politically correct" one, but can ideology struggle against scientific evidence? We decided that it was our duty to inform the public of the possible dangers of food carcinogens and to proceed with research.

The reaction from the media and from the scientific community was mixed. Some people simply did not believe the data, but later the experiments were repeated by laboratories all around the world, including at the Meat Research Institute. Some people said the data were not original, since similar results had been published by Sugimura and co-workers at the National Cancer Institute in Tokyo. We had actually quoted those results in our paper [1], which showed that mutagens were present in the charred parts of meat and fish, a process in which a lot of combustion compounds, including polycyclic aromatic hydrocarbons, are known to be formed [2]. A lot of scientists argued that it was not proven that mutagens present at low concentration in cooked meat were a health hazard. And after all, not all mutagens are carcinogens.

All these objections were disproved by later work done in this area, by us and other laboratories. Actually this debate prompted increased scientific interest in the field of food and cancer and scientific articles on this topic increased from a few per year to several hundred. Important and well-funded laboratories started investigating the "beef extract mutagens," which were later named "heterocyclic amines," since they had an amine group and contained several heterocyclic rings in their structure. Most of them were demonstrated to be carcinogenic for experimental animals, as suggested by our group. The source of the original observation was almost forgotten in later years and nearly everybody quoted the research of Sugimura's group in Tokyo, when referring to food heterocyclic compounds.

Later work in different parts of the world demonstrated that heterocyclic amines are invariably present in heat-processed meat and fish. It took about twenty years to realize that heterocyclic amines are probably involved in the induction of two of the most common cancers: breast and colon [3, 4].

We can now state with no fear of being contradicted that our original warnings were totally justified. Heterocyclic amines are important factors in human carcinogenesis: the effort to save the environment and defend people's health has to include better nutrition and reduced intake of carcinogens with food.

REFERENCES

1. B. Commoner et al., *Science, 201,* p. 913, 1978.
2. T. Matsumoto et al., *Mutation Research, 48,* p. 279, 1977.
3. E. G. Snyderwine, *Cancer, 74,* p. 1070, 1994.
4. P. Vineis et al., *Cancer Causes Control, 7,* p. 479, 1996.

CHAPTER 7

The Contribution of Barry Commoner to the Renewal of the Italian Left

Giovanni Berlinguer

Barry Commoner had a strong positive influence on the ideas and policies of the Italian left since the 1970s. His books were translated, and he was frequently present in the environmental movement. His criticism of the "Soviet model" found a favorable echo in the Italian Communist Party, whose autonomy had grown in the post-war period, and helped to include new ideas in its policy. Nowadays, left or center-left parties and alliances, often including "green" forces, lead the governments in thirteen out of fifteen countries belonging to the European Union, as a result of democratic elections. This is a result of both a strong tradition and of a renewal of strategies and programs to which Barry Commoner contributed with his ideas and his political courage.

I am very glad and honored to take part in this event on the past *and future* contributions of Barry Commoner to the environmental movement. Once Barry told me, joking, that he had had more books published in Italy than in the United States. Even if it is not true, the anecdote gives an idea of his impact on Italian culture and politics. I will try to analyze why and how it happened.

The first reason is the courage of his ideas and of his political behavior. He spoke at many universities and popular clubs in Italy. He visited several places to advise about the environmental impact of industries, and came to Seveso immediately after the ecological disaster produced by dioxin, to suggest what to do and how to avoid further pollution. He was the main speaker in the meeting called by the Istituto Gramsci on environment and economic development, which defined the strategy of the left on the issue. He was present as an activist guest in the Congress of the Italian Communist Party (I stress the word Italian) where an anti-nuclear policy was adopted. He even participated in my personal campaign for the Senate, and contributed to convince part of my electorate. I am very sorry, and apologize to Barry that I could not reciprocate this help during his campaign (for President as the candidate of the Citizens Party) in 1980. Don't worry: he did not give donations to interfere in the Italian elections. He did actually interfere by stressing and explaining the links between social and environmental problems, and the necessity of what can be called, with a simplified formula, a *red-green approach*.

IMPORTANCE OF OTHER IDEAS

The left in Italy—as in many countries in western Europe—is a wide, strong, working class-based movement, organized in unions and political parties, with a unique experience consisting of a strategy based on the conjunction of political democracy, free market, and social justice. At the same time, until the 1970s, the idea that the class struggle was the only moving force of history, and the only revolutionary movement, limited our understanding of society, our influence, and our renewal. Gradually, and sometimes through bitter struggles, the left understood the importance of other ideas and issues, mainly those raised by the women's movement and by the environmental movement, which grew particularly in the United States and had a positive impact on all Western countries. In Italy, probably more than elsewhere, because here the left was, because of the influence of Gramsci and the resistance against fascism, more and more heretical toward the Third International and Soviet dogmas and experience.

It welcomed, therefore, the analysis presented by Barry on the global situation and on the environment, and particularly the comments on the Soviet Union, presented at the meeting of the Istituto Gramsci in 1985. In his report, he said: "It is generally recognized that the Soviet society has failed to adequately develop the democratic process that is essential. What is perhaps less obvious is that the Soviet society has also failed to create technologies capable of generating mass production which, because they are deliberately designed for that purpose, serve the national interest and people's welfare." He concluded his contribution with the following words: "I believe that this task should be undertaken by the left in capitalist countries, and especially by political parties that have accepted the task of creating a democratic society."

I remember the comment I expressed to Barry: "You are probably too optimistic on the possibilities of the Italian left." In those years of the 1980s, my comment seemed to be confirmed by events. The collapse of the Soviet system and the triumph of the Reagan-Thatcher ideologies and policies put the left in Europe in a condition of struggling for survival. But the struggle was not only a defensive one. It was accompanied by the liberation of new ideas, and by a deep discussion on the perspective of the Western societies, mainly on the key question: Is it still possible to conjugate democracy, progress, and justice?

DISAGREEING WITH GLOTZ

The German social-democratic theoretician Glotz, for instance, affirmed that we live in a society where two-thirds of the population is protected (the employed, skilled, organized, and the privileged), and one-third, the excluded, suffers as a result of the current right-wing policies. I cannot agree with this analysis.

According to it, no social change could occur in a democracy (and we do not want to experience any more changes against democracy), because the feelings of solidarity toward the poor and the excluded have an enormous moral value, but cannot have the same political weight as material interests. Actually, I may add, the majority of the working class, even if it has enough goods and sometimes enough money, suffers from poor health and a poor quality of life. Moreover, half of the population—the women—are not guaranteed their rights, which are legally recognized but not implemented, either at work or in everyday life. Almost nobody is guaranteed, can feel safe from different forms of violence and self-destruction, such as drug addiction, which are increasingly worrying the Western countries, particularly in urban areas. Last, but most important, nobody can be totally protected against the deterioration of the natural environment, and everybody may suffer (both materially and culturally) from the fact that present generations are wasting resources and destroying monuments and masterpieces left by our ancestors, and are building very little for posterity. All of these phenomena are connected with production and social reproduction, but are not limited to this sphere. They do not simply demand an updating of the traditional ideas of class struggle, but are new dimensions of the reality and demand new programs.

I have given an example of the debate which took place in the European left since Barry and his ideas began to influence our strategies and contributed to renew our policies. I know that recently many persons abroad (and in our countries) were surprised by the successes of the left in Europe in the last two years, after the fall of the Soviet system which led somebody to declare "the end of history." The center-left alliance won the elections in Italy, the Labour Party in the United Kingdom, perhaps the socialist party will win in France, and huge strikes took place, for the first time in decades, in Germany.

I can see three explanations for this trend. One is a widespread reaction to the idea that the market economy can solve by itself all social problems, and a growing impatience toward the dominance, almost a cultural dictatorship, of monetary fundamentalism. The other is that the fall of the Soviet system liberated intellectual and political forces, and hence made each party, movement, group or person more free and responsible to choose his own way. Finally, new political alliances have been formed which include the left, part of the political center, and the greens.

EUROPE LEANS LEFT

At present, left-wing or center-left forces lead the governments of twelve (13 if the socialist party wins in France) of fifteen countries belonging to the European Union. Will it change the perspectives of the European societies in the fields of social justice, education, jobs, environment, and human rights, and the European commitment toward other parts of the world? Will it change the nature of the European Union? Originally, it was created as a "Steel and Coal Community," then became a Common Market, and now is dominated by the idea of a common currency. Its transformation will not be easy, because some of the problems faced by the European nations, like the budgetary deficit, must be solved, and there is a common interest in reducing inflation. But what about the deficit of living conditions and the environment, and the inflation of polluting activities, goods, and habits?

Let me say a few words about an example concerning the struggle against disease. The emphasis, in the scientific and even more in the popular literature, is now on lifestyles. In the *Decalogue against cancer*, formulated by the European Union, citizens are called to fully respect safety regulations during the production of carcinogenic materials, while it would be better to forbid the production and use of such materials. In a leaflet, it was even suggested that "when there is dangerous smoke in your city, stay at home." Probably Barry would have presented to people the opposite advice: when there is smog, go out into the streets, and ask the municipality and the state to clean the air; you will breathe worse for one day, but probably better for your whole life.

I mentioned this example because I am convinced that environment, health, education, and social justice are the new frontiers of the left; it would be more correct to say "of human societies," because these issues concern the present, and even more, future generations. Many thanks, Barry, for what you did to help us choose this way.

CHAPTER 8
Barry Commoner's Day

Chicco Testa

"Barry Commoner's Day" represents for me a great opportunity to settle up my debt of gratitude to Barry Commoner. I thank Barry for the personal friendship he has honored me with for many years, allowing me to take advantage of his experience, his good advice, and his scientific and political teaching. He used to be an incorrigible optimist. And I hope he has not changed with the passing of the years because God knows how much environmentalists need people like him since they have a tendency to complain and foretell misfortunes. But Barry's contribution is not just optimism. His great contribution lies in his ability in matching economic and social rationality, technological progress, and the minimization of environmental impacts. That is to say, the finding that at the basis of the processes involving the destruction of natural resources, there is often an irrational behavior, which is technologically and economically disadvantageous.

I must thank the organizers of this congress. It represents for me a great opportunity to settle up, even if partially, my personal debt of gratitude to Barry Commoner. The elements of this debt are quite a few. Some of them are very personal and of such a nature not to require a large public audience. In short, I must thank Barry first of all for the personal friendship he has honored me with for many years, allowing me to take advantage of his expertise, his good advice, and his scientific and political teaching.

Today, as some of you may know, I am the Chairman of the Board of Directors of ENEL, the Italian utility for the production, transmission, and distribution of electricity. It is the third largest electric utility in the world in terms of installed capacity, number of users, and income.

The day I was appointed by the new Italian government, an Italian newspaper wrote: "It's just as if the Apache chief became the President of the United States." In this sentence there was, of course, a typical journalistic exaggeration—Chicco

Testa does not look like an Indian warrior—but in this exaggeration there is also a little truth.

In particular, I suppose the journalist was referring to the fact that in the past, and not too far past, I was for many years the president of an Italian environmental association, Legambiente, and to the fact that this association had fought many battles against ENEL, among which the most famous one ended with a referendum which stopped the use of nuclear energy in Italy. Legambiente became in the space of a decade probably the most influential environmental Italian association, and counts among its founders Giovanni Berlinguer and Barry Commoner.

AN INFLUENTIAL BOOK

Before becoming President of Legambiente, I didn't personally know Barry, but I had read some of his books, in particular, *The Closing Circle,* which was included by a far-seeing professor among the texts for an exam at the Faculty of Philosophy in Milan, where I studied. This book was a revelation for me, and since it was translated into some dozens of languages and read by millions of people in the world, it is useless to tell you about it. But I can assure you that for many of us it represented a complete change of perspective. In substance and in short, this book was about the inclusion of the capital represented by natural and environmental resources into the assessment of the economic growth rates of our societies and into the assessment of the impact of different technologies, especially the energy ones, on these factors.

Today, many of these considerations seem obvious. Not because they have lost their importance but because, fortunately and at least in part, they have become a common concern and subject of many studies, research, and proposals. In some cases, they have also involved important changes in government and industry strategies, as well as in citizens' behavior. In many other cases, it didn't happen, and this is exactly what remains to be done, and there is a lot to be done.

But let's go back to the history of those days. When I had the opportunity to meet Commoner, it was an honor for me. Barry already had good connections in Italy, where he used to visit for many other reasons. I received an invitation to attend the Congress of the Citizens Party (in 1980) and it was on that occasion that our cooperation started, which brought Barry to play an active and decisive role in many Italian events, often as a leading actor.

In this connection, I would like to recall Barry's contribution to Italian energy policy, his studies on eutrophication of the Adriatic Sea and his proposals for a new strategy in the field of waste management. I don't know how much Barry cares, but I must inform him that, apart from the Italian decision to stop nuclear power, ratified by a popular referendum, which he knows well, Italy is undertaking an important program aimed at developing renewable energy sources, among which are almost 1000 megawatts of wind energy. Also, I am going to inaugurate soon the biggest photovoltaic plant in the world with about 3 megawatts of capacity. Moreover, for a long time, eutrophication has not appeared in the Adriatic Sea, in my opinion also thanks to the sharp reduction in nutrients which, following several laws, have been eliminated from many agricultural products and from industrial detergents. And, finally, I want to tell Barry that a few months ago Italy ratified an act requiring each municipality to implement waste recycling. Indeed, part of our program has been put into practice.

A FUNDAMENTAL CONTRIBUTION

But now I want to shift the focus of my speech from the chronicle and the history of the last years, to what I consider Barry's fundamental contribution to modern environmental culture. Those who know Barry well probably won't believe it, but some Italian journalists—I don't know if it is the same in the United States—still consider Barry a doomsday ecologist. In other words, an advocate of misfortune. I can affirm that Commoner is very different from this description and that, on the contrary, his thought and—with his consent— his character go right in the opposite direction. To tell the truth, he is an incorrigible optimist. And I hope he has not changed with the passing of the years, because I am an optimist too, and God knows how much the environmentalists need people like him, since they have a tendency toward complaining and foretelling misfortunes. This is the attitude used by children to call their parents' attention, but doesn't suit adults, who instead should use reason and should demonstrate the technological and economical feasibility of certain programs. Reason is not enough, of course. You need strength too, the pacific strength typical of democracy; but this latter also needs consensus, and you must wonder why whenever consensus lacks, you must not think that the others are always wrong.

Some of my environmentalist friends instead act like the character of an Italian joke which goes like this:

> A man goes out, takes his car, but by mistake he turns down the highway on the wrong side. He turns on the radio and the speaker says: "Attention! Attention! There is a madman on the highway who's driving in the wrong direction!" The man thinks: "Not only one madman, there are thousands!"

For some people, the word always goes in the wrong direction, and only they know the right way. And if ever somebody joins them, they start thinking about changing direction again.

But Barry's contribution is not just optimism. I would even say that his greatest contribution lies in his ability for matching economic and social rationality, technological progress, and minimization of environmental impacts. That is to say, the finding that at the basis of the processes involving the destruction of natural resources, there is very often an irrational behavior, which is technologically and economically disadvantageous. And that this irrationality, far from deriving from ignorance or lack of information, is the result—on the contrary—of the conflict of interests dividing advanced societies. And, as a consequence, the analysis and the construction of alternative options should be based on reliability, economic advantage, and feasibility.

This is a very innovative and reasonable approach in its premises and in its proposals.

DO NOT UNDERESTIMATE OPTIMISM

But the matter of optimism must not be underestimated, since it seems to me that the way of thinking of Commoner is based on trust—neither blind nor irrational, or course, but founded on analytical data. It is based on the innate capability of humans to improve their lot and to create real progress. That is to say, to eliminate the obstacles to improvement.

Let me mention a thinker, now out of fashion, Karl Marx, and his interpretation of the relation between the potential development of production forces and the obstacles raised against it by old feudal social relations in the society of that time.

Today we can affirm the same if we think about the obstacles still opposing development exerted by social and political relations which bridle the enormous potential of a far-sighted technological development. Barry Commoner is a man of our time, a modern man. In his environmental philosophy, there is no regret for the past, nor for the good old days. On several occasions, I had the opportunity to talk with him about the ideas expressed in the well-known report of the Club of Rome entitled, "The Limits to Growth," and afterwards, of some environmental philosophies derived from those ideas, philosophies based on a great distrust in the possibility of progress, enriched with doomsday and millenarian spirit, generally considering Nature like a divinity, the master of our actions. Sometimes Nature is even idealized as the bearer of moral values and rules. I want to say that I was somehow influenced too by some of those ideas: in particular by the book of the Club of Rome.

But I have to say now that I think Barry is right.

The distance between the time scale of natural processes and historic scales makes the two parameters not comparable. Better: there is no absolute and metaphysical limit to our actions which cannot be overcome by a rational use of the resources that we have at our disposal. This doesn't mean at all an absolute and boundless supremacy of human beings.

THE POSSIBILITIES IN OUR HANDS

An Italian poet-philosopher of extraordinary importance, Giacomo Leopardi, defined nature as a "stepmother," for the promises she makes when she gives us life and for how violently she takes everything back. But history and human societies have their own rhythms, included in the very long time expanse of natural evolution, and have possibilities which are all in our hands.

There is also a second point that I deem fundamental which was also stressed by Barry many years ago. It is the strict link between the environmental problem and the social problem. This means the end of hypocrisy, which tried to separate them, restricting the environmental problems within the circle of a few elites who pretended not to see this link. Sustainability, which today is a largely accepted principle, is not a neutral concept. It is not, as some environmentalists affirm, a sacrifice that humanity should endure to guarantee the respect of natural laws. The target, which Barry indicated much before the United Nations' report, is instead the finding of a balanced approach to development, where both factors, humans and nature, have their own advantage.

For most of the world's populations, these are concepts used by the rich nations and are completely incomprehensible. What meaning has the idea of solidarity with future generations for hundreds of millions of people who have a dramatic and often unsolvable problem of daily survival? Nature, in the form of dramatic natural unsolved daily problems, expresses its violence every day: with hunger, sickness, death.

The compromise that should be pursued must be based, therefore, on a cooperation which generates progress, based on the expectation of improvement, where a better future is entrusted to a new technological hope.

Personally, I am very thankful to Commoner for the friendship and the teaching of these years. Thanks to him, I have also started to know a part of American culture. Among other things, he also taught me the rules of baseball, and how to eat hot dogs and drink beer. These are important things in life, which must not be forgotten.

CHAPTER 9
What is Yet to Be Done

Barry Commoner

The environmental crisis expresses the relation between science and society in a special way: it illustrates the overriding importance of action. Action-oriented decisions—for example, whether to stop global warming for the sake of people or in order to conserve the natural world—profoundly affect the relationship. Both the post-World War II changes in production technology, which gave rise to the environmental crisis, and the failed effort to resolve it by the strategy of control, lead to a common conclusion: Environmental pollution is an incurable disease: it can only be prevented. By design, production technologies must be compatible with environmental quality. This introduces a social interest in what is widely regarded as a private prerogative: the decisions that determine what is produced and by what means. Environmental quality is then an aspect of political economy, requiring, for example, national, democratically determined, industrial and agricultural policies. Such a sweeping transformation of production can be powerfully inspired by a vision of the economic renaissance that would be generated by the new, more productive, technologies. The most meaningful engine of change may be not so much environmental quality as the economic development and growth generated by the effort to improve it.

When, in an excess of politeness, the organizing committee asked me whether I would like to speak at this wonderful symposium, I hesitated—uncharacteristically—because I knew that it would be hard to decide what to talk about. I should have known better, since, for its part, the committee—in keeping with an old Center for the Biology of Natural Systems (CBNS) tradition—did not hesitate to tell me what I ought to say. Perhaps in deference to my now undeniable seniority, the committee's instructions were delivered in a very tactful way, in the form of a four-word code disguised as the title of the symposium: Science and Social Action. Obviously my task was to extract, from only these four words, the committee's detailed instructions on how to make sense of the collective experience, wisdom, courage, and just plain hard work represented by all the good people here today.

In my attempt to break the committee's clever code, I was aided by the fact that the members, thoroughly indoctrinated in CBNS ideology, were likely to think in terms of certain environmental cliches, especially the main one: "Everything is connected to everything else." With that in mind, I quickly understood that the key word in the code was AND, signifying that science *and* society are closely connected, each dependent on the other. Surely I could find a lot to say about this meaty topic. That left the remaining code word: ACTION. This gave me the committee's basic operational instruction: "Don't overdo the philosophizing; just answer the question: 'What is to be done?'"

The interaction between science and social problems and the need to resolve them applies not only to the environment but to a number of other issues that also cry out for action, for example, health and—if we concede that economics is indeed a science—the economy. However, the environmental crisis is special, for it expresses the relation between science and society and the overriding importance of action in a distinctive way that illuminates the wider range of issues as well.

A FUNDAMENTAL FAULT

The environmental crisis arises from a fundamental fault: Our systems of production—in industry, agriculture, energy, and transportation—essential as they are, make people sick and die. As the Surgeon General would say, these processes are hazardous to your health. But that is only the *immediate* problem. Down the line, these same production processes threaten a series of global human catastrophes: higher temperatures; the seas rising to flood many of the world's cities; more frequent severe weather; and dangerous exposure to ultraviolet radiation. The non-human sectors of the living ecosystem are also affected by the crisis: ancient forest reserves are disappearing; wetlands and estuaries are impaired; numerous species are threatened with extinction.

Nevertheless, the environmental crisis is a *human* event; it is caused by what people do, and the ultimate measure of its impact is the health and well-being of people. I start with this assertion because all of us who profess to be environmentalists must decide for ourselves which of two alternative motives justifies environmental action: shall we stop the assault on the Earth's ecosphere for the sake of its human inhabitants—who depend on it—or to conserve the natural world itself? This is an unavoidable choice. For example, it determines whether or not action should be taken to avoid global warming. While, unchecked, the rising global temperature will surely devastate human society, the ecosystems that survive it would be no less "natural" than those of the earlier, equally warm carboniferous era. Only its human impact justifies the effort to prevent global warming.

The modern assault on the environment began about fifty years ago, during and immediately after World War II. I am grateful that my own adult life has covered this span of time, so that I have witnessed most of the notorious environmental blunders that led to the crisis—sometimes simply a bystander, other times an attentive observer, and at least once—in the case of DDT—an unwitting perpetrator. This experience, and my participation in the collective effort to understand and resolve the environmental crisis, has been enlightening. That effort has been marked by major environmental victories and disheartening defeats, and we can learn a good deal from both.

ENTERING THE ARENA

As you have heard, my own entry into the environmental arena was through the issue that so dramatically—and destructively—demonstrates the link between science and social action: nuclear weapons. The weapons were conceived and created by a small band of physicists and chemists; they remain a cataclysmic threat to the whole of human society and the natural environment. World War II had hardly ended when—not satisfied with the wartime bombs that killed hundreds of thousands of people in Japan—the United States and the Soviet Union began testing new and nastier ones, creating enormous amounts of radioactivity that spread through the air worldwide, descending as fallout. Many atomic scientists—alarmed by the consequences of their wartime work—protested. But the tests continued and were even expanded.

The tests were done in secret, marked only by Atomic Energy Commission (AEC) announcements that the emitted radiation was confined to the test area and, in any case, "harmless." This convenient conclusion reflected the AEC's assumption that the radioactive debris would remain aloft in the stratosphere for years, allowing time for much of the radioactivity to decay. But the AEC was wrong; in 1953, shortly after a test explosion, university physicists in Troy, New York, detected significant levels of its radioactivity in rainfall. At about the same time, a then-secret AEC report had concluded that no one was in danger from the radioactive strontium-90 in fallout unless they happened to eat a stray chip of bone in their hamburger. This, too, was wrong: as some of us in St. Louis pointed out, strontium-90 tracks calcium through the ecological food chain, from fallout-contaminated grass, into cows, their milk, and then—because milk, not bone, is their main source of calcium—into children's bones as they grow. Another AEC report concluded that fallout radioactivity levels were too low to "cause a detectable increase in mutations." Again the AEC was wrong; a United Nations scientific committee found that bomb tests would cause up to 100,000 serious genetic defects worldwide.

After 1954 when some of the secret reports were declassified, independent scientists were able to further analyze the fallout data that AEC scientists had developed but had failed to understand. The new analyses confirmed that they had grossly underestimated the dangers: E. B. Lewis, a geneticist at CalTech, showed that iodine 131, a major fallout component, was likely to cause thyroid tumors in children; Linus Pauling, the noted chemist, added carbon 14 to the roster of fallout hazards; Norman Bauer, a chemist at Utah State University, and E. W. Pfeiffer, a University of Montana zoologist, showed that there were high local fallout concentrations near, but outside, the Nevada test site; Erville Graham, a Canadian botanist, showed that the extraordinary capacity of lichens to absorb fallout directly from the air greatly amplified the hazards to native peoples in the Arctic.

FIELD-OF-VISION PROBLEM

The AEC had at its command an army of highly skilled scientists. Although they knew how to design and build nuclear bombs, somehow it escaped their notice that rainfall washes suspended material out of the air; or that children drink milk and concentrate iodine in their growing thyroids. Many people wondered whether the AEC did, in fact, know the truth, but suppressed it in an effort to avoid unfavorable publicity. Of course, this may have occurred. However, I believe that the main reason for the AEC's failure is less complex but equally devastating. The AEC scientists were so narrowly focused on arming the United States for nuclear war that they failed to perceive facts—even widely known ones—that were outside their limited field of vision. As a report from a Pentagon consultant pointed out as late as 1968, the ecological point of view "... has been strongly neglected ... and detailed research is conspicuously absent."

The AEC taught us that when science is forced to serve a powerful self-justified purpose, it becomes too narrow to serve the wider needs of society. It was the independent scientists, outside the AEC, who understood their obligation to society; it was they who met society's need for the truth. But despite their truth-telling and protests, the bomb tests continued and increased in number and intensity.

When the Committee for Nuclear Information (CNI) was organized in St. Louis in 1958, we did something different: we brought scientists *and* civic-minded citizens together. Our task was to explain to the public—in St. Louis and, soon, nationally—how splitting a few pounds of atoms could turn something as mild as milk into a devastating global poison. At about that time, several of us met with Linus Pauling in St. Louis and together drafted the petition, eventually signed by thousands of scientists worldwide, that is credited with persuading President Kennedy to propose the 1963 Nuclear Test Ban Treaty—the first of continuing international actions to fully cage the nuclear beast.

Yet, that did not mean victory, for the U.S. Senate was a nest of cold-warriors and, according to common wisdom, was unlikely to ratify the treaty. But the Senate was besieged by letters, many of them from parents who abhorred the idea of raising their children with radioactive fallout embedded in their bodies. What convinced the Senators was not so much their constituents' fear of radiation, but that they were *informed;* they knew how to spell strontium-90 and could explain precisely why it was so dangerous. The treaty was easily ratified.

The Nuclear Test Ban Treaty victory dramatized the political power of a public well informed by independent scientists about the technical facts and aroused to action by their own sense of the deadly threat—as close at hand as a glass of milk—of this man-made environmental contaminant. It was as much an environmental victory as a political one. It was an early indication of the collaborative strength of science and social action.

EXTENDING THE MISSION

It was this conclusion that led CNI to become the Committee for *Environmental* Information and extend its mission to the environmental crisis as a whole: the radiation dangers from nuclear power plants; the toxicity of DDT and other pesticides; the hazards of mercury, lead, PCBs and the huge amounts of noxious chemicals that the petrochemical industry produces—and disperses into the environment; the health effects of smog, which accompanied the auto industry's bloated post-war cars; the growing impact of chemical agriculture on water pollution.

These were man-made mistakes that were therefore within our power to remedy. The mistakes were made by the auto companies when they decided to build bigger cars with high compression engines that for the first time emitted nitrogen oxides, which in turn triggered the smog reaction; by the petrochemical industry that persuaded farmers to spread huge amounts of toxic pesticides—many of them carcinogenic—into the environment; by electric utilities that—believing government propaganda that nuclear power would be "too cheap to meter"—built the plants that generate highly radioactive spent fuel, which is yet to be dealt with. The sharp rise in environmental pollution in the twenty years following World War II could be traced to such new technologies of production—new ways of producing electric power, transportation, and food that, while they generated these valuable goods, now violently assaulted the environment as well. The changes were massive and fast: in less than two decades the total amount of automotive horsepower increased four-fold, of inorganic fertilizer nitrogen seven-fold, of synthetic organic chemicals twenty-fold.

In every case, the environmental hazards were made known only by independent scientists who were often bitterly opposed by the corporations responsible for the hazards. It was Rachel Carson, whose brilliant writing made public the toxic impact of DDT on wildlife—for which she was viciously attacked by the chemical industry. Arlie Haagen-Smit, a CalTech plant physiologist, discovered the cause of photochemical smog—a decisive result long denied by the auto industry. Charles Komanoff, among others (including Rob Scott at CBNS), predicted that nuclear power would be uneconomic—a conclusion ridiculed by the industry until it turned out to be so true that no plants have been built in the last twenty years. The occurrence of dioxin in Agent Orange that the U.S. Air Force sprayed on Vietnam was established by a Harvard microbiologist, Matthew Meselson, and its impact on the Vietnamese people documented by Dr. Arnold Schecter of SUNY Binghamton. The persistent work of Dr. Herbert Needleman, Professor of Pediatrics at the University of Pittsburgh, has established the devastating impact of environmental lead on children's intellectual development. Theo Colborn of the World Wildlife Fund and her colleagues have led the way toward an appreciation of the dangerous effects of endocrine disrupters. And I am proud that CBNS has played a role as well: to demonstrate that the huge post-war increase in the use of inorganic nitrogen fertilizer created serious water pollution problems; that large-scale organic agriculture can earn the farmer as much as conventional agriculture; and that—contrary to industry dogma—trash-burning incinerators actually synthesize dioxin as they operate, and are more costly than recycling.

IN APPRENTICES' HANDS

Of course, other scientists and organizations have made their contributions as well. I offer this selected list only to support two important generalizations about the role of science and scientists in generating what we know about the grave environmental impact of the post-war production technologies: First, the scientists, engineers and technologists who designed and built the new technologies—not to speak of their corporate masters—gave no public notice of their environmental faults because they were unaware of them, uninterested or, in some cases, deceitful. The vaunted sorcery of modern technology was hard at work—but environmentally, it was in the hands of apprentices. Second, outsiders were needed to set things right—or at least to help the American people learn what went wrong and why. As individuals, in ad hoc groups, or through more formal organizations, hundreds of scientists, often collaborating with local grassroots organizations, went to church groups, PTAs, and community organizations to explain how fallout, nuclear power, pesticides, and toxic dumps threaten the environment and the health and well-being of the people who live in it. The American people were informed, became concerned, and sought ways to act.

Those of us who have participated in the public debates about environmental issues have often marveled at the public participants' eagerness to learn. Armed with information provided by University of Alaska scientists, Eskimo villagers in Alaska learned enough about the distinctive biological behavior of strontium-90 in the Arctic to defeat the AEC's bizarre proposal to create a new harbor by exploding a hydrogen bomb; antinuclear activists in California became sufficiently acquainted with local geology to pinpoint the danger of an underwater fault near the site of a proposed nuclear power plant.

Earth Day 1970 was irrefutable evidence that the American people understood the environmental threat and wanted action to resolve it. The government quickly responded, and within the year, the National Environmental Policy Act (NEPA) established, as a national purpose, ". . . efforts that will prevent or eliminate damage to the environment." The Environmental Protection Agency (EPA) was created to administer these efforts, and beginning with the Clean Air Act, legislation was quickly enacted to establish specific remedial programs, encompassing the now-massive legislative and regulatory program, which extends into states and municipalities.

Environmental concern is now firmly embedded in public life: in education, medicine, and law; in journalism, literature, and art. It has turned hitherto indifferent politicians into self-proclaimed environmentalists, starting with Richard Nixon—an environmental non-starter who made the issue the centerpiece of his first State of the Union address. Looking back on these changes—or perhaps startled by the latest advertisement of an oil company that has turned itself green—we might be justified in proclaiming victory. Certainly, we have made things happen. But what has motivated environmentalism and, in my view, defines its purpose is the state of the environment itself.

FAR FROM VICTORY

By that measure we are far from victory: Neither the general aim stated in NEPA, nor the specific improvements mandated in the enabling legislation, have come even close to being achieved. The numerical evidence on the required improvements in air quality—which called for 90 percent reduction in pollution within seven years of 1970—is a persuasive example. According to the latest EPA assessment, after twenty-five years the best percentage improvement in emissions of the standard air pollutants (for sulfur dioxide) since 1970 is only 30 percent. Nitrogen oxide emissions have not improved at all over that period. Worse, in almost all cases whatever improvement did occur came to a halt after 1980; since then, except for a slow reduction in carbon monoxide emissions, the curves are flat. And EPA foresees no further improvement; its latest projections of air emissions show slight *increases* for all the standard pollutants from now to 2010, except for a small decrease in sulfur dioxide.

These numbers tell us that the methods that EPA introduced after 1970 to reduce air pollutant emissions worked for a while, but over time have become progressively less effective—and from the 1980s on no longer capable of reducing emissions at all.

The chief remedial method has been the installation of emission control systems—devices attached to the pollutant-generating source (such as autos, power plants, and incinerators) that trap and destroy the pollutants before they enter the environment. The fault is not that the control devices have themselves become less efficient since the 1980s. Rather, a countervailing process has overcome their emission-reducing capability. That process is economic growth: year by year, there are more cars and trucks on the road and more energy generated. As long as a control device is not perfect—that is, does not reduce emissions to zero—this increased activity counteracts the device's ability to reduce environmental pollution, and economic growth becomes the enemy of environmental quality.

This antagonism between the economy and the environment is built into the nature of a control device: for thermodynamic reasons, it becomes progressively more difficult—and therefore more costly—to remove the pollutant as the percent removed rises. For example, controls that remove 70 percent of sulfur dioxide from a coal-burning power plant's flue gas cost $50 per kilowatt of plant capacity; a system that removes 90 percent costs $2,000 per kilowatt, and it would take $4,000 per kilowatt to reach 99 percent—at which point the control system would cost ten times the cost of the power plant itself. As a result, it is simply economically impossible to require controls that even approach zero emissions. In turn, this economic limitation renders the control system vulnerable to the countervailing effect of increased economic activity. By adopting the control strategy, the nation's environmental program has *created* a built-in antagonism between environmental quality and economic growth.

THE FATAL EMBRACE

Tragically, this conflict—as well as the accompanying failure to meet the legislative goals of environmental improvement—could have been avoided if the enabling legislation had required EPA to abide by NEPA's stated purpose to *prevent and eliminate* pollution. By any interpretation, this requirement means *zero* emissions, which, if accomplished, would meet the mandated goals and undo the fatal embrace between the environment and the economy.

Ironically, hidden in the otherwise dismal data on air pollution emission trends, we can find concrete evidence that the strategy of prevention can actually achieve this astounding result. In 1970 U.S. vehicular transportation emitted 180,000 tons of lead into the air; by 1994, emissions had decreased to 1,600 tons—by 99 percent. This was achieved while vehicular transportation—a major economic activity—increased by 50 percent, as measured by fuel consumption. Environmental quality was drastically improved while economic activity grew by the simple expedient of removing lead from gasoline—which *prevented* it from entering the environment. There are a few similar examples of success by prevention in the environmental data. Environmental levels of DDT have decreased sharply—because its use has been banned; in some rivers, phosphate levels have declined a great deal—because phosphate-containing detergents have been banned. Meanwhile, neither agriculture nor detergent production has suffered.

These only-too-rare miracles have been accomplished by a well-known industrial practice: the technology of production—of gasoline, detergents, and cotton (the chief agricultural use of DDT)—has been altered, albeit at the behest of the government.

The task, then, is to apply the principle of pollution prevention to the major production processes that, in their present form, generate the mass of environmental pollution. There are existing pollution-free alternatives to the production technologies that brought on the post-war environmental crisis. The major source of photochemical smog—petroleum-fueled vehicles—can be replaced by emission-free electric vehicles. In turn, power plants now fueled by oil, natural gas, or uranium can be replaced by zero-emission photovoltaic cells or wind generators. The pollution-free alternative to current agricultural practice—which is heavily based on inorganic nitrogen and synthetic pesticides, both major causes of environmental pollution—is organic farming. In specific process industries, for example, paper mills, closed-loop systems that produce no effluents or emissions are feasible—one is already operating in Finland.

The new production technologies may be more economical than the ones they replace. For example, a recent CBNS study shows that the impact of trash-burning incinerators in the states adjacent to the Great Lakes on the airborne dioxin deposited in the lakes can be reduced to zero by diverting the trash to intensive recycling programs. The net economic effect would be a $500-million reduction in annual disposal costs, including the cost of paying off the incinerators' existing debt. As noted earlier, we have also found that large-scale organic agriculture is competitive with conventional practice in the Midwest.

WHY NOT THIS COURSE?

Why hasn't EPA pursued this course? Until 1989, pollution prevention was entirely absent from the EPA program, despite NEPA's assertion that it was the purpose of the national environmental effort. Then in January of that year, Lee M. Thomas, the retiring Administrator of EPA, published a "Pollution Prevention Policy Statement" in the Federal Register declaring that control measures had failed to satisfactorily improve the environment and that only prevention can succeed.

Although Mr. Thomas' departing testament sought to introduce pollution prevention as the EPA's guiding *policy* (that word, after all, appears in the title) in preference to pollution control, in practice EPA has reduced it to a subsidiary, rather than guiding, feature of the national program. Perhaps most indicative of its fate is that Mr. Thomas' portentous Pollution Prevention Policy has been given a cute acronym: P2. This even eliminates the word *policy* from the official acronym, which of course ought to be: P3. Meanwhile, where policy is actually expressed in EPA operations—in the promulgation of regulatory measures—pollution control holds sway. For example, in the latest regulatory document on dioxin emissions (for hazardous waste incineration), EPA prescribes a Maximum Achievable Control Technology.

Unlike EPA, direct public pressure, often organized by grassroots environmental and community organizations, *has* accomplished major pollution prevention measures. DDT fell victim to Rachel Carson's widely read *Silent Spring;* PCBs were banned in response to public outcry over a flurry of accidental releases. More recent examples include: the abandonment by McDonald's of plastic ware largely as a result of a children's campaign ("McToxic") organized by the Citizens Clearinghouse for Hazardous Waste; in many cities, trash-burning incinerator projects have been abandoned in favor of recycling under pressure from local grassroots groups; a consortium of such groups is waging a vigorous campaign to replace hospital waste incinerators with dioxin-free autoclaves.

These efforts are enormously important, for they provide concrete examples of how the abstract idea of "transforming the technology of production" can be turned into reality. But in keeping with the organizing committee's encoded admonition, I must add that there is a great deal yet to be done, beyond these local grassroots victories.

What is needed is a transformation of the major systems of production more profound than even the sweeping post-World War II changes in production technology. Restoring environmental quality means substituting solar sources of energy for fossil and nuclear fuels; substituting electric motors for the internal combustion engine; substituting organic farming for chemical agriculture; expanding the use of durable, renewable, and recyclable materials—metals, glass, wood, paper—in place of the petrochemical products that have massively displaced them.

RELATING ACTION, CORPORATE DECISIONS

In the U.S. economy, the decisions that determine what is produced and by what means are in private, generally corporate, hands. How can the demand for action to improve the quality of the environment, which is deeply embedded in society as a whole, be brought to bear on these private, corporate decisions?

I believe that the first step is to extend the environmental issue into the relevant social, economic, and political arenas. Consider, for example, the decision to replace conventional cars and light trucks with electric vehicles—powered, ultimately, from solar sources. The relevant corporations are reluctant to make this change because, compared with conventional ones, electric vehicles would initially be more costly and more restricted in their uses. Such a shift would damage a corporation's economic interests, they argue, in comparison with firms that refrained from making the change. However, this issue can be dealt with by establishing, as a national *industrial policy,* that all suitable vehicles are to be powered by electricity, placing all of the auto industry's firms on the same level playing field, economically.

There is nothing new about national policies on major social interests such as education or labor—or, for that matter, the environment. After all, despite the economic advantage to firms that employed child labor, it was in the *social* interest, as a national policy, to abolish it—removing that advantage for *all* firms. What is new is that environmentalism intensely illuminates the need to confront the corporate domain at its most powerful and guarded point—the exclusive right to govern the systems of production.

How can the environmental movement come to grips with this deeply rooted privilege that has firmly resisted even public discussion, let alone a proposed diminution? A useful way to approach this question is to think about it directly in economic, rather than environmental, terms. Seen that way, the wholesale transformation of production technologies that is mandated by pollution prevention creates a new surge of economic development. But this would touch on other social concerns as well. The wave of new productive enterprises would provide opportunities to remedy the unjust distribution of environmental hazards among economic classes and racial and ethnic communities. For labor unions, it would represent a source of new jobs and opportunities to advance the cause of a healthy work environment and worker retraining. Indeed, the transformation, although environmentally mandated, may be much more powerfully inspired by the vision of an economic renaissance that would be generated by the new more productive technologies. The most meaningful engine of change, powerful enough to confront corporate power, may be not so much environmental quality, as the economic development and growth associated with the effort to improve it.

A PREVAILING MYTH

Why should environmental advocates be in favor of economic growth, when this seems to fly in the face of the argument, often advanced by some environmentalists, that high rates of production and consumption are the chief *cause* of environmental degradation? That view is based on the assumption that production is *necessarily* accompanied by pollution, so that these two processes rise and fall together. It reflects a prevailing myth that production technology—the high-compression engine, the nuclear reactor, or genetic engineering—is simply the practical application of scientific knowledge and is therefore no more amenable to human judgment or social interests than the laws of thermodynamics, atomic structure, or biological inheritance. The environmental experience has shattered this myth. The high-compression engine and the nuclear reactor were built in response to *human* decisions and their linkage to smog and radioactive waste can be readily broken by building electric vehicles and photovoltaic cells instead.

On the other hand, there are powerful reasons why environmental advocates should favor economic development and growth—*that is based on ecologically benign technologies of production.* The most cogent reason is that the massive transformation of our major systems of production—that is essential to environmental quality—cannot achieve this goal if it is pursued only in developed countries.

The environmental crisis is a global problem and only global action will resolve it. Even concerted action by the northern industrialized countries—where most of the assault on the ecosphere now originates—will not be enough. What is done in developing countries is crucial as well. As the world population rises—90 percent of the increase in developing countries—worldwide production levels will need to increase sharply in order to sustain global economic development. Unless the expanded production facilities are ecologically sound, this process will further degrade the environment.

There are serious constraints on developing countries that, if unrelieved, will greatly reduce their ability to participate in the transition to ecologically sound systems of production. Since for some time the required production facilities—for example, solar energy equipment—would need to be imported, developing countries are potentially a huge market for the new environmentally benign products. In the United States and other developed countries, this demand would hasten the development of the transition and facilitate the growth of the new production facilities.

THE POWERFUL ENEMY

Of course, none of this can happen if we accept the false idea that environmental quality cannot tolerate economic growth. If, as I believe, the purpose of the environmental effort is to improve the health and well-being of people, then we must recognize that the most powerful enemy of human welfare is poverty. And we must remember that the human inhabitants of the Earth's ecosphere are engulfed in a global epidemic of poverty, hunger, and despair. The grim statistics can be summarized in a simple image. As the earth spins through space, a view from above the North Pole would encompass most of the wealth of the world—most of its food, productive machines, doctors, engineers, and teachers. A view from the opposite pole would encompass most of the world's poor. The planet is split by a chasm that separates the North from the South, the rich from the poor.

This global chasm *must* be bridged. This is the rational, logical outcome of the environmental experience. But I say to you that if environmentalism is to be devoted to human welfare, there are reasons more powerful than the environmental ones. Simple morality dictates that the rich should share their productive capacity with the poor. And an even more compelling imperative is justice, for the poor half of the planet has been brought to that plight through the exploitation of its resources and its people by the imperial nations of the North. We, who are environmental advocates, must find a way—for the sake of the planet and the people who live on it—to join a historic mission to end poverty wherever it exists. That is what is yet to be done.

Contributors

GIOVANNI BERLINGUER, a physician, was professor of Occupational Health at the University of Rome. He served in the Chamber of Deputies, the lower house of the Italian parliament from 1972 to 1983, and as a Senator from 1983 to 1992. He is currently the Director of the post-graduate School of Bioethics at the University of Rome. His fax is 0039 6 49912771.

VIRGINIA WARNER BRODINE is a writer of environmental journalism including the books *Air Pollution* and *Radioactive Contamination* (Harcourt Brace Jovanovich, 1973, 1975) and of historical fiction, including *Seed of Fire* (International Publishers, 1996). She was on the staff of the International Ladies' Garment Workers Union, AFL-CIO 1954-62 and editor of *Environment* magazine 1962-69. Her fax is 1-509-649-2223.

BARRY COMMONER is Director of the Center for the Biology of Natural Systems (CBNS), a research institute dedicated to finding solutions to environmental problems. Dr. Commoner founded the Center in 1966 at Washington University in St. Louis; in 1981 he moved the Center to Queens College, City University of New York. Contact him at CBNS, Queens College, Flushing, New York 11367.

PIERO DOLARA was affiliated with the Center for the Biology of Natural Systems as a Fogarty Fellow in 1997-98. He is now Professor of Toxicology at the University of Florence, Italy.

DANNY KOHL is Professor of Biology at Washington University in St. Louis. He had the pleasure of collaborating with Prof. Commoner on a number of scientific investigations over a long period of time. Contact him in the Department of Biology, Washington University, St. Louis, Missouri 63130-4899. His E-mail is Kohl@biodec.wustl.edu.

DAVID KRIEBEL is a Professor in the Department of Work Environment at the University of Massachusetts Lowell, where he co-directs the Lowell Center for Sustainable Production. He received his doctorate in epidemiology from the Harvard School of Public Health in 1986. His research interests include occupational and environmental causes of cancer and lung diseases. Professor Kriebel was a student of Dr. Commoner's beginning in 1972, and has worked closely with him ever since. His e-mail address: David_Kriebel@uml.edu.

CONTRIBUTORS

TONY MAZZOCCHI is a special assistant to the president of the Oil, Chemical and Atomic Workers international union and former vice president of the union. A long-time advocate of workplace health and safety, he also is a founder of *New Solutions* and of the Labor Party. His fax is 1-202-234-5176.

PETER MONTAGUE is the direct of the Environmental Research Foundation in Annapolis, Maryland and editor of the foundation's newsletter, *Rachel's Environment and Health Weekly.* Telephone: 1-410-263-1584; E-mail: peter@Rachel.clark.net.

RALPH NADER, an attorney, is considered the father of the consumer movement and is one of America's most respected public advocates. He founded the Center for Study of Responsive Law and Public Citizen, whose membership exceeds 100,000 and whose member organizations include Congress Watch, Health Resarch Group, Critical Mass Energy Project, Global Trade Watch, and the Litigation Group. Nader also has authored many books. His fax is 1-212-234-5176.

CHICCO TESTA was born in Bergamo. In 1980, he was appointed National Secretary of the Italian Environmental League; he later became its Chairman. He has served twice in the Chamber of Deputies, and became chief of the Party of the Democratic Left's group within the Italian Parliamentary Commission for the Environment. He chairs the Board of Directors of ENEL, the Italian national electric utility. His fax is 011 39 6 850 97954.

Index

a-disciplinary science, 2
AAAS, *see* American Association for the Advancement of Science
above-ground weapons testing, 9
acceptable risk, 13
addiction, 34
adenosine triphosphate, 47
Adriatic Sea, 68, 69
AEC, *see* Atomic Energy Commission
Agent Orange, 78
Air Pollution Foundation, 32
air pollution, 21
American Association for the Advancement of Science, 46
American Tort Reform Association, 32
Ames, Bruce, 57, 58
Ames mutagenesis assay, 57
anemia, 34
asbestos, 37
assimilative capacity, 10
Atomic Energy Commission, 17, 19, 26, 46, 75, 78
atomic fallout, 6, 7, 9, 45
ATP, *see* adenosine triphosphate

Baby Tooth Survey, St. Louis, 8, 9, 18, 25, 28
baby teeth, 9
bacterial mutagenesis assay, 58
Bauer, Norman, 76
Bauer, Walter, 16
Baumgarten, Judy, 16
beef extract, 58
Bennett, Jim, 48
Berlinguer, Giovanni, 68
Bikini test, 26
Bingham, Dr. Eula, 29
Bodega Bay, 20
bone, 75
Brodine, Virginia, 27
Brown and Williamson Tobacco Co., 33, 34
burden of proof, 10

^{14}C, *see* carbon-14
calcium, 75
cancer, 33
carbon-14, 76
carboniferous era, 75
carcinogenesis, 51, 60
Carson, Rachel, 15, 77, 82
Castleman, Barry, 38
CBNS, *see* Center for the Biology of Natural Systems
CEI, *see* Committee for Environmental Information
Center for Public Integrity, 36
Center for the Biology of Natural Systems, 1, 57, 58, 73, 78
center-left alliances, 65
Central Dogma of Molecular Biology, 47
Chamber of Commerce, U.S., 38
champagne, 53
child labor, 83
Citizens Party, 64, 68
Citizens' Clearinghouse for Hazardous Waste, 82
civil defense, 19
Clear Air Act, 79
Closing Circle, The, 16, 52
Club of Rome, 70
CNI, *see* Committee for Nuclear Information
Codex Alimentarius, 42
Colburn, Theo, 78
Committee for Environmental Information, 15, 77
Committee for Nuclear Information, 5, 15, 16, 25, 27, 76
Committee for Sane Nuclear Policy, 27
corporate front groups, 41
corporate power, 83
corporate science, 32
cotton, 81

DDT, 75, 77, 80, 82
Decalogue against cancer, European, 66

Democratic Party, 25
Department of Defense, U.S., 19
developing countries, 84
dioxin, 63, 78
diphtheria-tetanus-pertussis vaccine, 33
DNA, 45, 47, 48
dogma, 47
DPT vaccine, 33
Drug Amendment Act of 1962, 35
Dubos, Rene, 27

Earth Day, 79
earthquake, 20
East St. Louis, IL, 21
ecosphere, 84
Eisenhower, President Dwight D., 8
electric utility, 67
electron spin resonance, 51
emission controls, 79
emphysema, 33
ENEL, 67
Environment Magazine, 17
Environmental Protection Agency, 79, 80
environmental impact assessment, 12
environmental movement, 15, 64
environmental pollution, *see* environmental quality
environmental quality, 73, 77
EPA, *see* Environmental Protection Agency
Erlich, Paul, 29
ESR, *see* electron spin resonance
European left, 65
European Union, 63, 66
eutrophication, 69

fallout, nuclear, 19
fallout shelters, 19
FDA, *see* Food and Drug Administration
fertilizer, 53
fertilizer nitrogen, 45
fertilizer runoff, 53
Food and Drug Administration, U.S., 34, 35
food chain, 75
food safety, 42
fossil fuel emissions, 36
Fowler, John, 16
Franklin, Rosalind, 46
free radicals, 6, 50-52

Gaffron, H., 51
Galileo's Revenge, 32
gasoline, reformulated, 42
GATT, *see* General Agreement on Tariffs and Trade

Gelhorn, Edna, 16
General Agreement on Tariffs and Trade, 31, 41
General Motors Corp., 37
global warming, 35, 36, 73, 74
Glotz, 65
Gordon, Gloria, 16
Gosling, R.G., 46
Graham, Erville, 76
grass roots environmental movement, 5, 7, 17, 78
Great Lakes, 81
guanosine triphosphate, 48

H-bomb testing, 9
Haagen-Smit, Arlie, 32, 78
hamburgers, 59, 60
heart disease, 33
Heise, John, 51
heterocyclic amines, 60
Hoechst, 34
Huber, Peter, 32, 35, 38
hydrogen bomb, 9

ILGWU, 16
incinerators, 1, 78, 81
industrial policy, 82
International Ladies' Garment Workers Union, 16
iodine-131, 76
Istituto Gramsci, 63, 64
Italian Communist Party, 63
Italian left, 64

Japan, 75
junk science, 31-33, 35, 36

Karsh, Robert, 21
Kennedy, President John F., 19, 76
Kessler, David, 34
kidney damage, 34, 35
Kilgore, U.S. Senator Harvey, 46
Komanoff, Charles, 78

Labor Party, British, 65
labor unions, *see* trade unions
Latham and Watkins, 40
Lawrence Livermore Laboratories, 40
lead, 77, 80
Lederle Laboratories, 33
Legambiente, 68
Leopardi, Giacomo, 70
Levitan, Judith, 26
Lewis, E.B., 76

lichens, 76
Limits to Growth, The, 70
litigation, obstruction of, 39
Los Angeles, 33
lung disease, 33

magnetic field, 50
malaria, 45
Manhattan Institute, 32, 41
market economy, 65
Marx, Karl, 70
McCaull, Julian, 22
McNeil Laboratories, 34, 35
Meat Research Institute, 60
Meecham, Stuart, 27
mercury, 77
Merital, 34
Meselson, Matthew, 78
Michl, Leo, 52
milk, 9, 16, 75
Molecular Biology of the Gene, The, 47, 48
Monnet, Jean, 42
Montague, Peter, 38
moral wisdom, 8
mutagens, 60
mutation, 58

^{14}N, 53
^{15}N, 53
N, 45
N fertilizer, 53
Nadler, Al, 27
NAFTA, *see* North American Free Trade Agreement
National Academy of Sciences, 19
National Environmental Policy Act, 12, 79, 80
Needleman, Herbert, 78
NEPA, *see* National Environmental Policy Act
neurological illness, 33
New Solutions, 30
New York Committee Against Nuclear Weapons Testing, 27
New York Public Interest Research Group, 41
nitrate, 53, 54
nitrogen, 45
nitrogen oxides, 77, 79
Nixon, President Richard M., 79
No Contest, Corporate Lawyers and the Perversion of Justice in America, 40
North American Free Trade Agreement, 31, 41
Novick, Sheldon, 22

Nuclear Engineering Corporation, 40
Nuclear Information, 17
Nuclear Test Ban Treaty of 1963, 76, 77
nuclear energy, 17, 37
nuclear fallout, 6, 16, 26, 45, 75, 76
nuclear missiles, 39
nuclear power, 68, 77, 83
nuclear reactor, 83
nuclear war, 15
nuclear weapons testing, 1, 9, 16, 26-28

O'Connor, John, 36
obstruction of litigation, 39
Occupational Safety and Health Act, 25
Occupational Safety and Health Administration, 29
occupational health and safety, 25, 28, 29
oil industry, *see* petrochemical industry
Olson, Walter, 32
organic farming, 1
organic pollutants, 9
OSHA, *see* Occupational Safety and Health Administration
over-population, 29
oxidation, 49
oxidation-reduction reactions, 51
Ozonoff, David, 36

Pacific Gas & Electric Co., 20
Pake, George, 50
Partial Test Ban Treaty, 19
Pauling, Linus, 76
Paulson, Glen, 27
PCB, *see* polychlorinated biphenyls
pesticides, 1, 21
Peterson, Malcolm, 21
petrochemical industry, 29, 35, 77
Pfeiffer, E. W., 76
Philip Morris, 34
photochemical smog, 32, 77, 78, 81, 83
photovoltaic cells, 81
PIRG, New York, 41
Politics of Energy, The, 1
pollution, *see* environmental quality
pollution prevention, 10, 81
polychlorinated biphenyls, 58, 77, 82
population growth, 29
Poverty of Power, The, 1
precautionary principle, 6, 11
Price-Anderson Act, 38
production, 73
Project Chariot, 20
Project Harbor, 19
Project Plowshare, 20

Public Citizen Health Research Group, 34, 35

R. J. Reynolds, 34
radiation, nuclear, 9
Reagan, President Ronald, 38
Reagan-Thatcher ideology, 64
recycling, 1, 81
red-green alliance, 63
redox reactions, 51
reformulated gasoline, 42
Reiss, Dr. Louise, 18
Republican Party, 26
right to know, 6, 10
Rio de Janeiro Conference on Environment and Development, 11
Rio Declaration, 11
Risk Assessment, 6, 12, 13
RNA, 46, 48
runoff, 53

St. Louis Baby Tooth Survey, *see* Baby Tooth Survey
St. Louis, MO, 21
salmonella microsome test, 58
San Andreas Fault, 20
SANE, 27, 28
Sangamon River, 52-54
Schecter, Arnold, 78
science information movement, 1, 8
Scientist and Citizen, 17
Scientists' Institute for Public Information, 21
Scott, Robert, 78
seatbelts, 37
secrecy, antithetical to science, 7
sequestration theory, 47, 49
eveso, Italy, 63
Shea, Kevin, 22
Shearer, Georgia, 54
Shell Oil Co., 21, 22, 29
Silent Spring, 15, 82
SIPI, 21, 28
SLAPP suits, *see* Strategic Lawsuits Against Public Participation
smog, *see* photochemical smog
smoking, 33
Snoderly, Max, 27
social action, 22, 73, 74
social responsibility, 2
solar energy, 36-38, 81
sound science, 42
Soviet Union, 17, 63, 64, 75
spin, electron, 50

Stapp, John Paul, 37
Stevenson, Adlai, 25, 26
Stockholm, Sweden, 28
Stokes, A. R., 46
Stossel, John, 32
Strap Hangers Group, 41
Strategic Lawsuits Against Public Participation, 31, 40
strike, labor, 29
strontium-90, 9, 16, 18, 25, 28, 77
Sugimura, 60
suits, legal, 21, 22, 31, 39
sulfur dioxide, 79, 80
Suprol, 34
Surgeon General, U.S., 34
sustainability, 71
systems thinking, 2

Teamsters' Union, 54
technology, 73, 78
thalidomide, 35
Third International, 64
Thomas, Lee M., 81
threshold, toxic effects, 9
tobacco, 33, 34
tobacco mosaic virus, 46
Townsend, Jack, 50
Toxic Deception, 36
Toxics Release Inventory, 11
trade unions, 64, 83

U.S. Chamber of Commerce, 38
U.S. Ecology, 40
UNCED, 11
unions, *see* trade unions
universities, 39
university-industry joint ventures, 39

Vietnam war, 45

water pollution, 53
water supply, 11
Watson/Crick model, 46, 47
Waxman, U.S. Representative Henry, 34
Weiss, U.S. Representative Ted, 34
Who Owns the Sun?, 36
Wigner, Eugene, 19
Wilkins, M. H. F., 46
Wilson, H. R., 46
Winona, Texas, 40
working class, 27, 29, 64, 65
workplace safety, *see* occupational health and safety
World Trade Organization, 41